GEOTECHNICAL SPECIAL PUBLICATION NO. 109

EDUCATIONAL ISSUES IN GEOTECHNICAL ENGINEERING

PROCEEDINGS OF SESSIONS OF GEO-DENVER 2000

SPONSORED BY
The Geo-Institute of the American Society of Civil Engineers

August 5–8, 2000
Denver, Colorado

EDITED BY
Norman D. Dennis, Jr.

1801 ALEXANDER BELL DRIVE
RESTON, VIRGINIA 20191–4400

Abstract: These proceedings contain eight papers that address a wide variety of issues in engineering education. Specific focus is given to the incorporation of newer technologies into geotechnical engineering classes such as the integration of multimedia courseware and virtual laboratories as well as innovative applications of traditional concepts like design integration and experiential preparation. Issues relating to pedagogy and the early attraction of students to the engineering profession are interspersed throughout the proceeding.

Library of Congress Cataloging-in-Publication Data

Geo-Denver 2000 (2000 : Denver, Colo.)
Educational issues in geotechnical engineering : proceedings of sessions of Geo-Denver 2000 : August 5-8, 2000, Denver, Colo. / sponsored by the Geo-Institute of the American Society of Civil Engineers ; edited by Norman D. Dennis, Jr.
p. cm. – (Geotechnical special publication ; no. 109)
Includes bibliographical references and index.
ISBN 0-7844-0522-0
1. Engineering—Study and teaching—Congresses. 2. Engineering geology—Study and teaching—Congresses. I. Dennis, Norman D. II. American Society of Civil Engineers. Geo-Institute. III. Title. IV. Series.

T62 .G46 2000
624'.071'1--dc21 00-042142

Any statements expressed in these materials are those of the individual authors and do not necessarily represent the views of ASCE, which takes no responsibility for any statement made herein. No reference made in this publication to any specific method, product, process or service constitutes or implies an endorsement, recommendation, or warranty thereof by ASCE. The materials are for general information only and do not represent a standard of ASCE, nor are they intended as a reference in purchase specifications, contracts, regulations, statutes, or any other legal document. ASCE makes no representation or warranty of any kind, whether express or implied, concerning the accuracy, completeness, suitability, or utility of any information, apparatus, product, or process discussed in this publication, and assumes no liability therefore. This information should not be used without first securing competent advice with respect to its suitability for any general or specific application. Anyone utilizing this information assumes all liability arising from such use, including but not limited to infringement of any patent or patents.
Photocopies: Authorization to photocopy material for internal or personal use under circumstances not falling within the fair use provisions of the Copyright Act is granted by ASCE to libraries and other users registered with the Copyright Clearance Center (CCC) Transactional Reporting Service, provided that the base fee of $8.00 per article plus $.50 per page is paid directly to CCC, 222 Rosewood Drive, Danvers, MA 01923. The identification for ASCE Books is 0-7844-0522-0/00/ $8.00 + $.50 per page. Requests for special permission or bulk copying should be addressed to Permissions & Copyright Dept., ASCE.

Copyright © 2000 by the American Society of Civil Engineers, All Rights Reserved.
Library of Congress Catalog Card No: 00-042142
ISBN 0-7844-0522-0
Manufactured in the United States of America.

Geotechnical Special Publications

1 *Terzaghi Lectures*
2 *Geotechnical Aspects of Stiff and Hard Clays*
3 *Landslide Dams: Processes, Risk, and Mitigation*
4 *Tiebacks for Bulkheads*
5 *Settlement of Shallow Foundation on Cohesionless Soils: Design and Performance*
6 *Use of In Situ Tests in Geotechnical Engineering*
7 *Timber Bulkheads*
8 *Foundations for Transmission Line Towers*
9 *Foundations & Excavations in Decomposed Rock of the Piedmont Province*
10 *Engineering Aspects of Soil Erosion, Dispersive Clays and Loess*
11 *Dynamic Response of Pile Foundations–Experiment, Analysis and Observation*
12 *Soil Improvement: A Ten Year Update*
13 *Geotechnical Practice for Solid Waste Disposal '87*
14 *Geotechnical Aspects of Karst Terrains*
15 *Measured Performance Shallow Foundations*
16 *Special Topics in Foundations*
17 *Soil Properties Evaluation from Centrifugal Models*
18 *Geosynthetics for Soil Improvement*
19 *Mine Induced Subsidence: Effects on Engineered Structures*
20 *Earthquake Engineering & Soil Dynamics II*
21 *Hydraulic Fill Structures*
22 *Foundation Engineering*
23 *Predicted and Observed Axial Behavior of Piles*
24 *Resilient Moduli of Soils: Laboratory Conditions*
25 *Design and Performance of Earth Retaining Structures*
26 *Waste Containment Systems: Construction, Regulation, and Performance*
27 *Geotechnical Engineering Congress*
28 *Detection of and Construction at the Soil/Rock Interface*
29 *Recent Advances in Instrumentation, Data Acquisition and Testing in Soil Dynamics*
30 *Grouting, Soil Improvement and Geosynthetics*
31 *Stability and Performance of Slopes and Embankments II*
32 *Embankment of Dams–James L. Sherard Contributions*
33 *Excavation and Support for the Urban Infrastructure*
34 *Piles Under Dynamic Loads*
35 *Geotechnical Practice in Dam Rehabilitation*
36 *Fly Ash for Soil Improvement*
37 *Advances in Site Characterization: Data Acquisition, Data Management and Data Interpretation*
38 *Design and Performance of Deep Foundations: Piles and Piers in Soil and Soft Rock*
39 *Unsaturated Soils*
40 *Vertical and Horizontal Deformations of Foundations and Embankments*
41 *Predicted and Measured Behavior of Five Spread Footings on Sand*
42 *Serviceability of Earth Retaining Structures*
43 *Fracture Mechanics Applied to Geotechnical Engineering*
44 *Ground Failures Under Seismic Conditions*
45 *In Situ Deep Soil Improvement*
46 *Geoenvironment 2000*
47 *Geo-Environmental Issues Facing the Americas*
48 *Soil Suction Applications in Geotechnical Engineering*
49 *Soil Improvement for Earthquake Hazard Mitigation*
50 *Foundation Upgrading and Repair for Infrastructure Improvement*
51 *Performance of Deep Foundations Under Seismic Loading*
52 *Landslides Under Static and Dynamic Conditions–Analysis, Monitoring, and Mitigation*
53 *Landfill Closures–Environmental Protection and Land Recovery*
54 *Earthquake Design and Performance of Solid Waste Landfills*
55 *Earthquake-Induced Movements and Seismic Remediation of Existing Foundations and Abutments*
56 *Static and Dynamic Properties of Gravelly Soils*
57 *Verification of Geotechnical Grouting*
58 *Uncertainty in the Geologic Environment*
59 *Engineered Contaminated Soils and Interaction of Soil Geomembranes*
60 *Analysis and Design of Retaining Structures Against Earthquakes*

61 *Measuring and Modeling Time Dependent Soil Behavior*
62 *Case Histories of Geophysics Applied to Civil Engineering and Public Policy*
63 *Design with Residual Materials: Geotechnical and Construction Considerations*
64 *Observation and Modeling in Numerical Analysis and Model Tests in Dynamic Soil-Structure Interaction Problems*
65 *Dredging and Management of Dredged Material*
66 *Grouting: Compaction, Remediation and Testing*
67 *Spatial Analysis in Soil Dynamics and Earthquake Engineering*
68 *Unsaturated Soil Engineering Practice*
69 *Ground Improvement, Ground Reinforcement, Ground Treatment: Developments 1987-1997*
70 *Seismic Analysis and Design for Soil-Pile-Structure Interactions*
71 *In Situ Remediation of the Geoenvironment*
72 *Degradation of Natural Building Stone*
73 *Innovative Design and Construction for Foundations and Substructures Subject to Freezing and Frost*
74 *Guidelines of Engineering Practice for Braced and Tied-Back Excavations*
75 *Geotechnical Earthquake Engineering and Soil Dynamics III*
76 *Geosynthetics in Foundation Reinforcement and Erosion Control Systems*
77 *Stability of Natural Slopes in the Coastal Plain*
78 *Filtration and Drainage in Geotechnical/Geoenvironmental Engineering*
79 *Recycled Materials in Geotechnical Applications*
80 *Grouts and Grouting: A Potpourri of Projects*
81 *Soil Improvement for Big Digs*
82 *Risk-Based Corrective Action and Brownfields Restorations*
83 *Design and Construction of Earth Retaining Systems*
84 *Effects of Construction on Structures*
85 *Application of Geotechnical Principles in Pavement Engineering*
86 *Big Digs Around the World*
87 *Jacked Tunnel Design and Construction*
88 *Analysis, Design, Construction, and Testing of Deep Foundations*
89 *Recent Advances in the Characterization of Transportation Geo-Materials*
90 *Geo-Engineering for Underground Facilities*
91 *Special Geotechnical Testing: Central Artery/Tunnel Project in Boston, Massachusetts*
92 *Behavioral Characteristics of Residual Soils*
93 *National Geotechnical Experimentation Sites*
94 *Performance Confirmation of Constructed Geotechnical Facilities*
95 *Soil-Cement and Other Construction Practices in Geotechnical Engineering*
96 *Numerical Methods in Geotechnical Engineering: Recent Developments*
97 *Innovations and Applications in Geotechnical Site Characterization*
98 *Pavement Subgrade, Unbound Materials, and Nondestructive Testing*
99 *Advances in Unsaturated Geotechnics*
100 *New Technologies and Design Developments in Deep Foundations*
101 *Slope Stability 2000*
102 *Trends in Rock Mechanics*
103 *Advances in Transportation and Geoenvironmental Systems Using Geosynthetics*
104 *Advances in Grouting and Ground Modification*
105 *Environmental Geotechnics*
106 *Geotechnical Measurements: Lab & Field*
107 *Soil Dynamics and Liquefaction 2000*
108 *Use of Geophysical Methods in Construction*
109 *Educational Issues in Geotechnical Engineering*

Preface

Educational activities, especially at the undergraduate level, often take a back seat to other scholarly activities, much to the detriment of our profession. Compounding this problem is the need for educators to incorporate and ever expanding body of knowledge into programs with decreasing numbers of credit hours. In order to provide quality education under these constraints educators of the new millennium will have to take advantage of new and innovative technologies available for use in the classroom and laboratory.

The objective of this publication is to provide the reader with information on recent innovations in the presentation of geotechnical engineering education. Special emphasis is placed on the incorporation of newer technologies into geotechnical engineering classes such as the integration of multi-media courseware and virtual laboratories as well as innovative applications of traditional concepts like design integration and experiential preparation. Issues relating to pedagogy and the early attraction of students to the engineering profession are interspersed throughout the proceeding

This publication was developed under the sponsorship of the Geo-Institute of ASCE. The papers that it contains were presented at GeoDenver 2000, August 5–8, 2000, in Denver, Colorado. In accordance with the practice of the Geotechnical and Geoenvironmental Engineering Division, each paper published in a special publication is reviewed for content and quality. Each paper in this volume has received at least two peer reviews, and authors were given the opportunity to modify their papers to address reviewer comments. The Editor wishes to acknowledge and thank the following members and friends of the Geo-Institute who assisted in the process of reviewing the papers:

Brian R. Brenner
Roy F. Cook
Norbert J. Delatte
Norman D. Dennis
Ro DiBrezzo
David Druss
Lewis Edgers
Amanda Elioff
Mark D. Evans
S. Trent Parkhill

All published papers are eligible for discussion in the Journal of Geotechnical and Geoenvironmental Engineering, and are also eligible for ASCE awards.

Editor:

Norman D. Dennis, Jr.
University of Arkansas
Fayetteville, AR

Abstract

Engineering educators in the new millennium will face varied and difficult challenges in providing quality instruction to their students. Effectively conveying an expanding knowledge base into programs that have ever-diminishing credit hours is just one example of the challenges that face the academic profession in general and civil engineering education specifically.

In order to meet these challenges educators of tomorrow must be armed with the latest innovations in teaching technology and pedagogy. These proceedings contain eight papers that address a wide variety of issues in engineering education. Specific focus is given to the incorporation of newer technologies into geotechnical engineering classes such as the integration of multi-media courseware and virtual laboratories as well as innovative applications of traditional concepts like design integration and experiential preparation. Issues relating to pedagogy and the early attraction of students to the engineering profession are interspersed throughout the proceeding.

Contents

Experiential Preparation for the Classroom

Ronald B. Meade, M.ASCE[1]
Christine M. Fiori, A.M.ASCE[1]
Mark S. Malone, A.M.ASCE[1]

Abstract

The United States Air Force Academy (USAFA) has provided a comprehensive "hands on" introduction to the civil and environmental engineering curricula. The introductory course known as Civil Engineering Practices-Field Engineering is required of all rising civil and environmental engineering juniors. The three-week laboratory is offered each summer at the Field Engineering Readiness Laboratory (FERL). The philosophy underlying this course is that students benefit from a "hands-on" experience prior to introduction of the technical principles in the classroom. This idea is exemplified by the course motto, "construct first, design later."

The cadets perform a wide variety of engineering tasks during the course. This paper discusses the geotechnical activities performed at FERL and describes how they are integrated into the core classroom work that is required in the soil mechanics course (junior year) the pavements course (senior year) and the geotechnical senior design elective course.

The paper describes the soils walk, compaction testing, and the road building activities that take place during FERL. Examples are presented from both the cadet and instructor viewpoint showing how the FERL experience is linked to the classroom and traditional laboratory. Finally, the assessment process used to determine the effectiveness of the "construct first, design later" integration process is outlined.

Approach

Relevancy is one of the keys to successful instruction. The Air Force Academy believes that relevancy can be created by providing a "hands-on" experience for students who have just declared the civil or environmental engineering major. Their experience begins with a two-week visit to an operational Air Force Civil Engineering Squadron at several bases

[1] Assistant Professor, Department of Civil and Environmental Engineering, United States Air Force Academy, CO 80840

throughout the Pacific Command. Upon their return from the Pacific, the cadets get an intensive three-week course that introduces them to construction activities. Subsequent classroom courses in the junior and senior year can build on the construction experience when introducing elements of design. The motto for this approach is "Construct First, Design Later."

The execution of this approach must be carefully crafted to integrate the construction experience with concepts learned in the classroom. The integration effort takes place on two levels, physical and intellectual. The physical level is grounded in a field laboratory that we call the "Field Engineering Readiness Laboratory" or FERL. The intellectual level is found in homework assignments and class exercises that require the students to recall the construction experience as a setting for the classroom exercises.

This paper describes the integration employed in the geotechnical program at the Air Force Academy. Parallel efforts take place in structures, environmental, and construction practices courses within our curriculum. The cadets are aware that this integration is a vital feature of the engineering experience at the Academy.

Background

The concept of an integrated engineering education has been evolving at the Academy since 1992. Colonel David Swint, the department head, envisioned a curriculum that would be alive with relevancy and produce graduates that would have an unparalleled understanding of engineering concepts and practices. The physical and intellectual components of this curriculum needed a fully equipped field laboratory to support planned student activities that would provide support for later learning of concepts.

Inquiry began into which hands-on activities would translate well into the classroom allowing complete integration between application and theory. Additionally these activities should draw upon the expertise of the non-commissioned officer mentors and faculty. The talents and strengths of each civil engineering subspecialty (structures, environmental, geotechnical and construction) are highlighted in the CE351 curriculum and lend themselves to "hands-on" learning.

During the period 1992-1995, activities included surveying techniques, construction of homes designed by senior students, lab testing of construction materials such as wood, concrete and steel, roadway paving, and construction and design of a concrete beam and steel bridge.

In later years (1996-1999), environmental and geotechnical applications were added to enhance the evolving curriculum. They include water sampling from a local creek, open channel flow demonstrations, a 3-reservoir setup, completion of soil sampling, a Standard Penetration test, compaction testing and a soils walk.

FERL – The Site

The field laboratory was planned in 1992 and ground was broken in 1993. Today the site includes 22 acres and six buildings. The cadets live at the site during the three-week course in a "construction camp" experience.

Civil Engineering 351

The course that provides that construction experience is called Civil Engineering 351 "Civil Engineering Practices - Field Engineering." The course includes 17-training days and is held in June each year. All of the faculty members are involved in the course and the cadets are divided into 5 flights (or groups). The flights are lead by a cadet junior. The flights were formed in May and travel as a group to one of five Air Force Bases overseas to observed both flight and engineering operations at active bases. During the past two years the cadets spent two weeks at a base in the Pacific Command. The flights include Academy cadets and ROTC cadets from selected programs. Seventy-seven cadets from USAFA and 21 ROTC cadets participated last year (June 1999).

Geotechnical Activities

The Geotechnical division instructs three primary courses throughout the academic year. These include a basic soil mechanics course, an advanced soils design course addressing slope stability, bearing capacity and foundations, and a pavement design course. Traditionally, the geotechnical segment of CE351 included a roadway paving activity and a soils walk. This year the activities were expanded to include a drill rig demonstration, field compaction quality control tests and a compaction lesson. Table 1.0 outlines the geotechnical activities conducted during CE351.

GEOTECHNICAL ACTIVITIES AT FERL			
ROADWAY	**SITE INVESTIGATION**	**PLANT VISITS**	**COMPACTION TESTING**
Roadway layout	Hand auger profiling	Portand Cement Plant	Proctor Test
Subgrade compaction	Soil boring	Asphalt Plant	Sand Cone Density
Base course placement	Standard Penetration Testing		
Base course compaction			
Hot Mix Asphalt Placement			
Hot Mix Asphalt compaction			

Table 1.0 - Geotechnical Activities at FERL

Soils Walk

Students were lead on a walk through three different areas of the FERL site. Soil samples were taken with a hand auger and brought back to the soil laboratory for later use. During the outing, students learned soil field classification procedures and gained an understanding of the different soil types, their formation, transportation and engineering properties. Additionally, erosional processes were highlighted that impact the formation of local topography.

Drill Rig

A trailer-mounted drill rig was used to demonstrate soil sampling and testing methods. The students were shown boring with a solid-stemmed auger. They were asked to describe the cuttings. The Standard Penetration Test (SPT) was demonstrated and the cadets counted the blows to advance the spoon and determined the blow count. The cadets were shown the contents of the spoon.

Compaction Testing

This activity readily connected theory and practice. Earlier students placed and compacted base course for the hot mix asphalt placement activity, without any compaction specifications or quality control. Additionally, students had encountered the soil used for this activity at varying moisture contents during the heavy equipment operation phase of the curriculum held the previous week. These experiences were drawn upon to illustrate the importance of the moisture-unit weight relationships via the compaction test.

Discussion focused on the concerns of strength and settlement in using soil as a construction material. Specifically, how imparting mechanical energy and adding or subtracting water would affect the soil behavior. To tie theory to application, standard Proctor tests were completed to establish the maximum dry density of the soil and the optimum moisture content. Students were told how compaction specifications were developed based upon the compaction test results.

Sand-Cone Density Testing

Cadets were placed into groups of three and an upperclassman demonstrated the sand cone density test. Then, each group performed a test on their own. The students were told about the relationship between the compaction testing that establishes compaction specifications and the field density tests that determine that the specifications have been met.

Road-Building Activities

This is the first opportunity the cadets have to utilize their new surveying and equipment operating skills. From a surveying standpoint, the cadets are tasked with laying out the centerline of the roadway and staking the edges to determine appropriate roadway widths and thickness. Placement, grading and compaction of the base course material is then completed before the asphalt arrives on-site in dump trucks operated by the cadets. Cadets place about 2 tons of asphalt in a day, resulting in about 500 feet of two-lane roadway. These activities are illustrated in Figure 1.0. To add to the reality of the paving operations, cadets are also responsible for traffic control throughout the entire operation

Figure 1.0 - Road building activities at the FERL site

Associated Activities

Each cadet is given the opportunity to operate construction equipment. The students operate a bulldozer, a scraper, and a grader. During the road building activity, selected students operated a dump truck, a front-end loader, and rollers (vibratory smooth-wheeled for earth and smaller steel-wheeled for asphalt finishing). The soil used in the compaction activity was gathered from the soil that had been worked by the students during the scraper operation. This same soil was later used in the classroom activity in the soil mechanics course in the fall semester.

Geotechnical Courses

Civil Engineering 390 Soil Mechanics

The cadets will enter the basic soil mechanics course having just completed the CE351 curriculum and therefore have a basic understanding of the potential uses for soil as a construction material. To kick-off the course, cadets are taken to an area of campus near the academic buildings to complete a soils walk and to collect soil samples. The area they are exposed to in-class exhibits different types of soils and topography, but cadets can draw on their CE351 experience to develop ideas on how the soils may have been formed, how they were transported and how these soils may behave.

Soils collected by the cadets during the soils walk conducted at FERL are incorporated into the laboratory experiments completed in this class. Specifically, the sand material collected was used in the sieve analysis experiment and fine-grained clay materials were used for the Atterburg limits test. The students remember collecting these soils and enjoy the classification process, since they only guessed as to soil type before. The compaction laboratory was also conducted on soils from the FERL site. This tie allows the students to relate the soil to a specific location. It is material that they moved with a scraper, compacted with rollers and collected with the hand auger. It is not soil that just "appeared" in the lab for them to complete their tests.

Civil Engineering 488 Transportation of Pavements

The cadets enter the pavement course having built a section of hot mix asphalt roadway during CE 351. The asphalt road that runs to the FERL site is used as the subject for an exercise in geometric design of a highway. The students have driven that road and have observed its features (grade, width, drainage, and earthwork). The cadets are required to prepare a new geometric design to re-route a section of that road to avoid an environmentally sensitive area.

Civil Engineering 491 Geotechnical Applications

The cadets are taken on field trips during the class periods to the FERL site to observe a drill rig demonstration, examine a set of three retaining wall types, and evaluate earthwork. The FERL site is about 7 minutes driving time from the classroom. The field trips are easily made during a 50-minute period.

Selected cadets had the opportunity to drive an SPT sample and all of the class determined the blow count. The course includes the design of walls. The FERL site has three wall types expressly created for student observation. The site has a soldier-beam and lagging wall, a reinforced concrete cantilever wall and an MSE block wall. The students learn the design principles about retaining structures and they are shown a video that was made of the construction of those walls at FERL.

The cadets are provided with the design drawings for the FERL wall area and they are required to inspect the site and note the "as-built" conditions and contrast those conditions with the features shown on the design drawings. They are also required to make observations concerning the effectiveness of weep holes and site works, such as slopes and surface water drainage features.

Assessment

In order to ensure continuous improvement of the course and to validate the impact CE351 has on United States Air Force (USAF) civil engineering officers, intensive evaluation of the curriculum occurs every summer. These efforts include feedback focus groups, and informal feedback during the follow-on major courses.

The focus group feedback sessions are conducted by personnel from the USAFA Center for Educational Excellence (CEE). The CEE is used for two reasons. First of all, the CEE is staffed with people knowledgeable in the field of course development and evaluation. Course assessment is their area of expertise. And secondly, by utilizing personnel from outside the department to gather the feedback the cadets give more candid and honest responses.

Each of the five flights of approximately 18 students comprises a "focus group". This includes a mixture of USAFA cadets, ROTC students, and maybe a Coast Guard Academy cadet or USMA cadet. The students are asked to list the strengths and weaknesses of the course and then assign a rating to the course from 1 to 5, with 5 as the highest. They are also asked several open-ended questions that were written by the department. The conversations are recorded on audio-tape, but the cadets are identified only be a number. The tapes are transcribed and the results analyzed by the CEE. A sample of the results from the Summer 99 offering of CE 351 are included as Figure 2.0.

The responses to our survey enabled us to assess the success of our integration efforts. The answers we received ranged from very positive, in which the cadets were able to clearly see the importance of the curriculum integration to extremely negative, in which the cadets could draw no significant benefit from the integration attempt. A sample of positive and negative responses is presented in Table 2.0. However, the intent of the FERL experience is best captured in the following survey comment," All was good [Geotechnical activities]. Didn't really know what it was all about until this class [CE390] though. It's valuable after the fact. (looking back)."

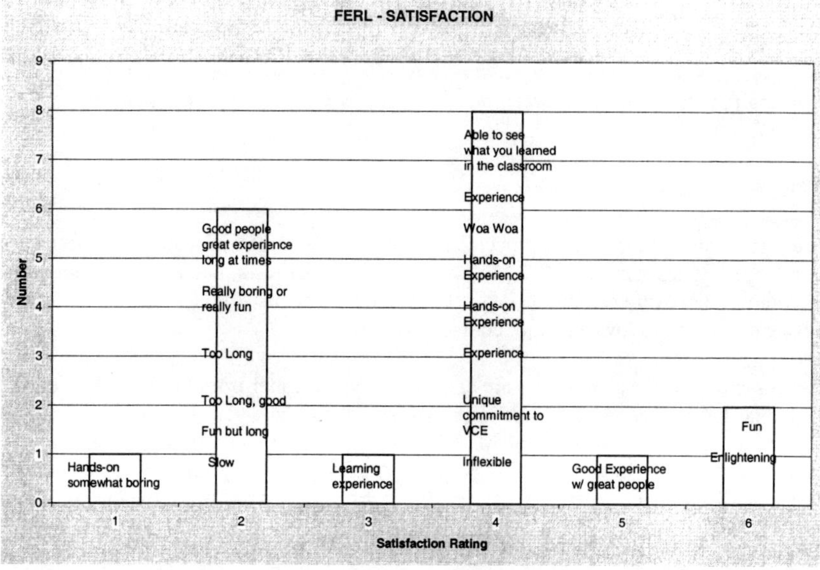

Figure 2.0 - FERL Focus Group Results

Using the sliding scale below, answer the following question.

What is the value in testing in the CE390 Laboratory the soils introduced at FERL?

1	2	3	4	5
Not Important		Somewhat Important		Very Important

Figure 3.0 - Sample assessment questionnaire

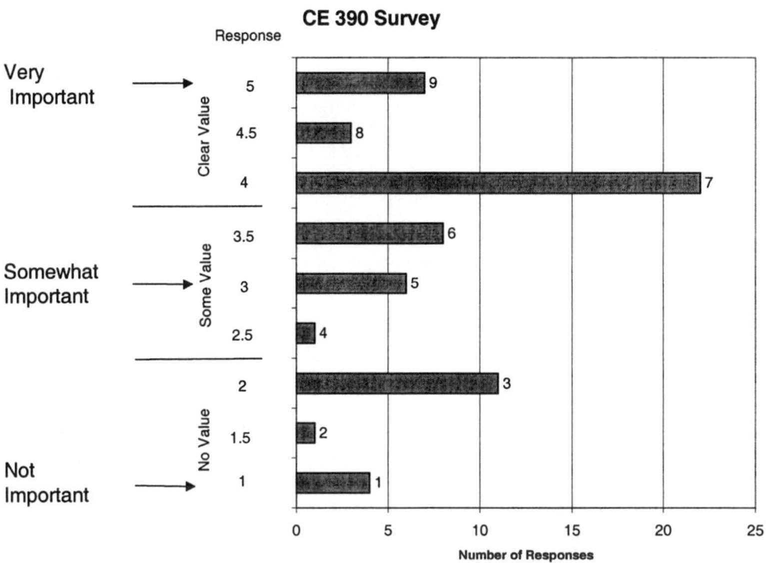

Figure 4.0 - Histogram of CE390 Assessment Survey

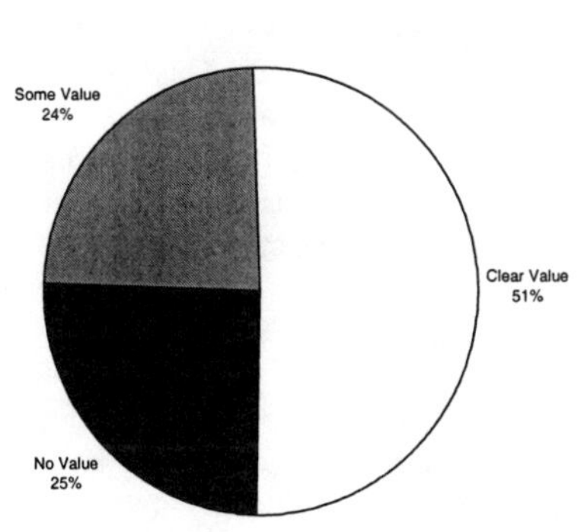

Figure 5 - Results of Soil Assessment Survey

POSITIVE	NEGATIVE
"I think FERL familiarized us with soil types and how to go about testing them. I know I learned that there is a lot more to soils than people think."	"None, because by the time we did the testing in the Lab I had no clue where we got the soil or what the terrain around looked like."
"I can visualize concepts learned in class having seen and touched them before. I can also understand better the importance of soils in construction having put down a road."	[testing the FERL soils was]"not really important, I could have taken the teachers word for it."
"FERL was like a real-world application to what we are learning in CE390."	"I can't say that knowing where the soil came from made much difference."

Table 2.0 - Sample positive and negative assessment survey feedback

Summary/Conclusions

We were encouraged by the results of our informal assessment survey regarding the integration efforts between the CE351 experience and CE390. Three-quarters of the cadets found value in testing the soils first introduced at FERL within the CE390 laboratory, a true testament to the success of the integration. However, the survey also highlighted areas that require some improvement. We plan to address these areas to concentrate on strengthening the link between FERL and the classroom by incorporate photographs from the site and a site map indicating where the soils were gathered. This information will give a visual reminder to the cadets as they begin their laboratory assignments. Additionally, all of the integration efforts have focused on CE390. We would also like to develop stronger field links to the CE491 design course and the CE488 Transportation and Pavements Course. Overall, the results showed that integration is worth the effort.

Acknowledgements

The authors would like to thank the Center for Educational Excellence for the use of their FERL focus group data within the paper. Additionally, we would also like to thank the cadets for their candid responses to our assessment surveys that allowed us to analyze the success of our integration efforts.

Integrated Geotechnical Design Experience

Mark D. Evans[1], M. ASCE and Steven J. Ressler[2], M. ASCE

Abstract

This paper describes an integrated soils and foundations course, soils laboratory, and capstone design course in which all cadets (students) were working toward a common goal of designing a "real-world" project. Site selection and project planning was kicked-off in the fall semester by a small group of seniors, who later acted as group leaders for the capstone design course in the spring semester. Juniors in the soils and foundations course got involved by excavating test pits, collecting and testing soils, and acting as consultants in designing foundations. Thus, the project was integrated into three courses over two semesters and included both junior and senior-level students. All cadets were involved in briefing academy officers and potential project donors leading to a $500,000 private donation.

Introduction

There is considerable emphasis in the literature on "service learning", learning by doing, and integrating "real-world" projects into capstone design courses (Albano and Salazar (1998), Anderson (1992), Gokhale and Aldrich (1997), Jones (2000), Lamancusa et al. (1997), McEwen (1994), Meyer and Jacobs (2000), Morris and LaBoube (1995), Pauschke and Ingraffea (1996), and Sender (1998)). In this paper we describe a project-based approach to learning design that integrates three courses and students of two year groups. A junior-level soils and foundations course was integrated with a senior-level capstone design course and a senior-level individual study course to produce a design for an actual project at the US Military Academy (USMA) in the spring term of 1999. This effort goes beyond the hands-on

[1] Associate Professor and Group Director, Department of Civil and Mechanical Engineering, US Military Academy, West Point, NY 10996.
[2] Professor and Civil Engineering Program Director, Department of Civil and Mechanical Engineering, US Military Academy, West Point, NY 10996.

geotechnical engineering approach described by Fiegel and DeNatale (1998), taking activities to the next level. Not only was this our first civil engineering (CE) capstone involving the design of an actual project for a real-world client, but it was also our first major CE project in which seniors and juniors worked together in three different courses to develop a single comprehensive design. The components of the integration are shown in Figure 1. The three courses involved in this integration were: CE492 - Design of Structural Systems (Capstone), CE371 - Soil Mechanics and Foundation Engineering (Soils), and CE489 - Advanced Individual Study. It should be noted that the Soils class is a one-semester, 3.5 credit-hour, 40-lesson course covering undergraduate soil mechanics and foundation engineering, with an associated soil mechanics laboratory consisting of eight 2-hour lab sessions.

This paper will describe the entire project and all course interaction, but will focus on the geotechnical aspects of the integrated design.

Project History and Description

West Point's climate has always put the Cadet Golf Team at a competitive disadvantage as the relatively long winters limit available practice time on the West Point golf course. In order to reduce the northern-climate disadvantage, the superintendent of the USMA expressed an interest in the design and construction of a training facility for the golf team. General requirements of the project were to provide the team with a state-of-the-art golf training facility to service indoor and outdoor training. The facility was to include an indoor putting area, indoor chipping green, club repair shop, training rooms, men's and woman's showers and lockers, bathrooms, laundry room, kitchen, meeting room, and coaches office.

The project began when three Civil Engineering (CE) faculty members learned that the proposed golf facility was on the "wish list" of the golf team coach and the USMA Alumni group (Association of Graduates). The facility was on the books as a potential gift-funded project, but the prospects for actual design and construction were poor, due to low priority and lack of prospective donors. The CE instructors offered to do preliminary design work for the building as a cadet project. The alumni group readily accepted, recognizing the marketing value of cadet involvement in the project.

Course Integration

Work began in the fall semester, as eight cadets in an Individual Study (CE489) conducted a site survey, defined the scope and functional requirements, developed a preliminary design, then briefed it to the golf team coach, the Athletic Director (Deputy Director of Intercollegiate Athletics), and project managers from the West Point Physical Plant (Directorate of Housing and Public Works) as shown in Figure 2. Based on feedback from this briefing, the facility was used as the capstone design project (CE492) during the spring term. Eight cadet teams each developed a

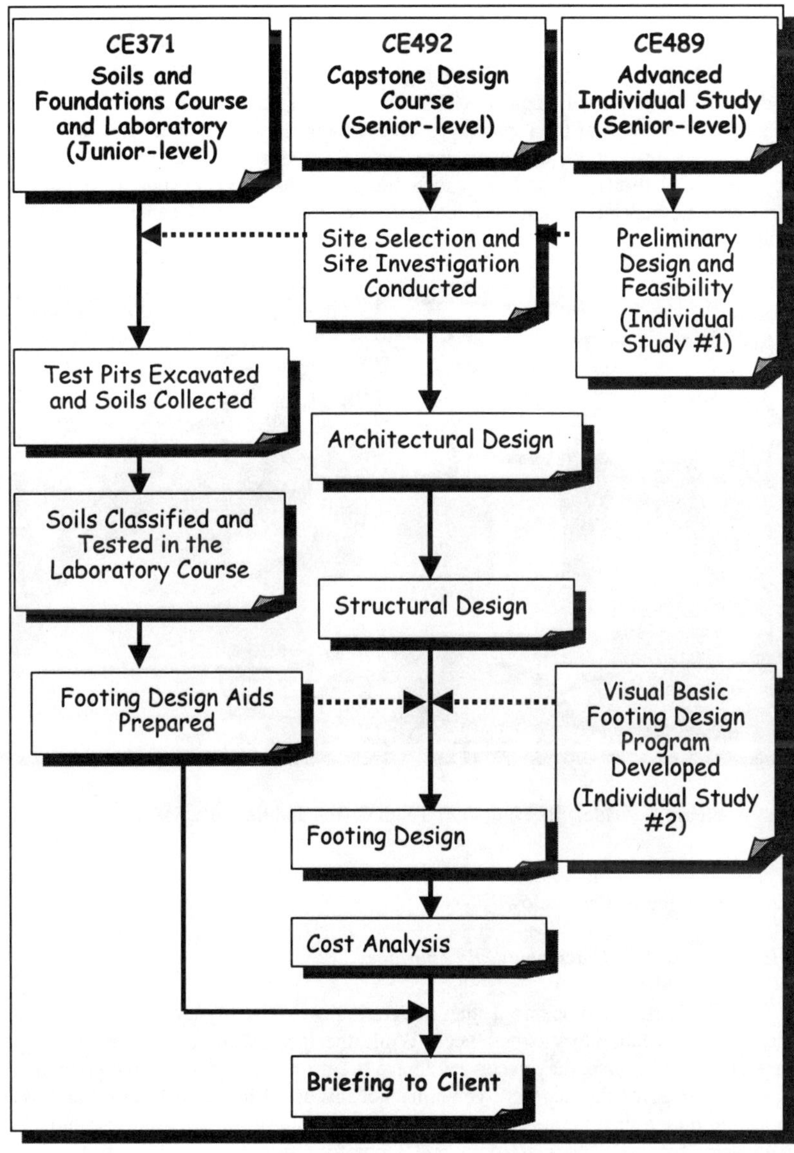

Figure 1. Components of the Integrated Geotechnical Design Experience

complete design for the facility, to include architectural, structural, foundation, and site designs, and detailed cost estimates (Allyn et al., 1999, unpublished). Meanwhile, the project was also integrated into the classroom and laboratory component of the Soils course (CE371). Soils students were briefed on the geotechnical aspects of the capstone project early in the soils course, and aspects of the capstone project were woven into the course and soils laboratory at every opportunity. Ultimately, the Soils cadets served as foundation design consultants to the Capstone cadets - a first-ever integration of two classes of cadets on a single project.

Figure 2. Cadets Briefing Golf Team Coach and Senior USMA

Individual components of the geotechnical coordination and integration are described in the sections below.

#1: Seamless Soils Classroom and Laboratory

The first component of the integration was the creation of a seamless, classroom and laboratory experience. While the lab and the course are packaged together in one course, they necessarily have been run as somewhat separate entities. Specific details of the lab were generally not discussed in the classroom, and vice versa, although instructors do take every opportunity to relate classroom and laboratory topics. The classroom and laboratory space were physically located adjacent to each other to facilitate maximum integration of both, allowing the instructor to move from classroom to lab and back as necessary. The course and lab lessons were also interwoven so that laboratory topics were covered immediately following their treatment in the classroom. Thus, portions of some labs, or lab demonstrations, were conducted during classroom hours, and similarly, some classroom lessons were covered during lab periods – generally two classroom lessons taught during one 2-hour lab period. Most of these structural changes were done specifically for this course and project integration, although they could have been implemented independently. Some of the structural changes continue today, and student comments on the course will be addressed later in this paper.

#2: Soils Collected and Tested by Students

The second component was that the soils tested in the laboratory by cadets were also collected by cadets in the field. USMA provided a backhoe and operator at the project site so that the cadets could log what was observed in the test-pit walls and make judgments about where to collect soil samples. Capstone students located the building onsite, and the soils students located three test pits and observed their excavation. Test pit excavation and cadet sampling is shown in Figure 3.

Figure 3. Test pit activities at the project site.

Cadets were involved in test pit sampling, and sufficient soil was collected and stored in 5-gallon buckets so that the cadets would retain those particular samples for the duration of the laboratory exercise. All laboratory testing for CE371 was conducted on those same soils in accordance with applicable ASTM (1999) standards. The junior cadets prepared professional reports and transmittal letters to pass soils data through their instructor to the senior cadets in CE492, following standards established in CE Standards for Technical Work (USMA, 1991). The emphasis was always placed on clear, thoughtful presentation of test and design data prepared in a professional manner (Evans, 1995). An abbreviated summary memorandum prepared by one cadet soils group is shown in Figure 4 (Melin and Workman, 1999, unpublished). It should be noted that the data in Figure 3 were used for preliminary design of the project – not for actual design. Professional engineers from outside USMA will be hired for final design and analysis.

#3: Soils from Real Project Site

As previously described, the third component was that the soils collected were from an actual project site for an actual "real-world" project of some interest to the cadets. This may have been the single most important aspect of this integration; bringing real life and meaning to the laboratory tests the cadets were conducting. It served as a focus for the instructor as well as he tried to get the cadets to reflect on the soils properties obtained, whether they made sense, and how they would affect the project. As each laboratory test was completed and results determined, students would occasionally repeat tests if the data was inconsistent with previously determined values. In addition, the test results were not so many "meaningless values" determined to simply fill a report, but were real data to be used for preliminary project design. Thus, slightly more care and consideration was apparent in the laboratory testing and discussion of results. Student comments regarding their laboratory tests are shown below:

Reasonableness of results:

1) *Soil Properties: Our results for the USCS classification of the soil match our results for visual identification. The liquid limit test and plastic limit test yield a negative plasticity index. Although this shows some error in measurement, it shows demonstrates the low plasticity of our soil. The total unit weight found using the sand cone test is extremely high for any soil. The porosity of the soil falls within the normal range for a sand; however, our saturation was fairly high. This may have been because our total unit weight was too high. Out graph of permeability vs. void ratio showed almost no linear relationship, which suggests that there was error in calculating at least one of the values.*

MEMORANDUM 12 May 1999

SUBJECT: Laboratory 10, Complete Soil Write-up

1. PURPOSE: The purpose of this report is to summarize the findings of laboratory test conducted on the soils at the West Point Golf Team Indoor Training Facility. Using this information, contractors will determine design parameters for the foundation of the facility.

2. RESULTS:

a. USCS Soil Classification: Found in accordance with ASTM D2487.

Soil 1 (to depth 24 in.): Well Graded Sand with Silt, *SW-SM* - 33% subangular gravel, 67% sand

Soil 2 (to depth 75 in.): Silty Sand with Gravel, *SM* - 32% subangular gravel, 48% fine to course subangular sand, 20% fines

b. Liquid Limit, Plastic Limit, Plasticity Index: Found in accordance with ASTM D4318.

Soil 1: Liquid Limit, LL = 22%; Plastic Limit, PL = 20 %; Plasticity Index, PI = 2%.

Soil 2: Liquid Limit, LL = 19-22%; Plastic Limit, PL = 22-24%; Plasticity Index, PI = 0

c. Moisture Content, Specific Gravity: In accordance with ASTM D2216, D854.

Soil 1: Moisture content, w = 23-25%

Soil 2: Moisture content, w = 8-10%; Specific Gravity, G_S = 2.7

d. Total Unit Weight: In accordance with ASTM D1556.

Soil 1: Saturated Unit Weight, γ_{sat} = 118-120 pcf

Soil 2: Saturated Unit Weight, γ_{sat} = 143-145 pcf.

e. Void Ratio, Porosity, Saturation, Permeability:

Soil 1: Void Ratio, e = 0.908;

Soil 2: Void Ratio, e = 0.30-.33 Porosity, n = 24%; Saturation, S = 85.4%; Permeability = 0.019 cm/s.

f. Compaction Test: In accordance with ASTM D698

Soil 1: Optimum Moisture Content, OMC = 13-15%; Max Dry Unit Weight, $\gamma_{d(\max)}$ = 110-115 pcf.

Soil 2: Optimum Moisture Content, OMC = 10-12%; Max Dry Unit Weight, $\gamma_{d(\max)}$ = 120-125 pcf.

g. Direct Shear Test:

Soil 1: Cohesion, c = 0 psi; Internal Friction Angle, $\phi = 40°$

Soil 2: Cohesion, c = 0 psi; Internal Friction Angle, $\phi = 38°$

h. Triaxial Test:

Soil 2: Cohesion, c = 0 psi; Internal Friction Angle, $\phi = 40°$

Soil 2: Cohesion, c = 0 psi; Internal Friction Angle, $\phi = 38°$

3. DISCUSSION:

a. Reasonableness of results: *Not included here.*

4. If you have any questions, feel free to contact CDT Melin at (914) 555-1234.

NICHOLAS MELIN CDT SGT, A-3 RYAN WORKMAN CDT SGT, F-1

Figure 4. Memorandum Transmitting Summary Laboratory Test Results

2) *Compaction: Our compaction curve is at best an approximation because we had only three usable data points. Our last point is also very close to the zero air void line, which leads us to believe it might be inaccurate.*

3) *Soil Strengths: We found that our Mohr's failure envelopes for the direct shear and triaxial tests were almost identical. Although our values confirmed that strength increases as moisture content decreases, we had at least one outlier. Our results for dry unit weight versus shear strength showed no real trend. We concluded that the direct shear and triaxial tests were good.*

Sources of error:

1) *Soil Properties: Most of the error in testing for soil properties can be attributed to disturbing the soil during excavation and preparing it for testing.*

2) *Compaction: The error in the compaction test came from the conduct of the lab itself. During the test we sometimes allowed the hammer to imbed itself in the soil or to strike the soil at an angle. We also failed to properly distribute the blows over the entire area of the soil. For the last test we added too much water, giving the soil the consistency of mud. This made compaction impossible.*

3) *Soil Strength: We found little error in our conduct of the triaxial test and direct shear test. Both used computerized systems, making the results very accurate. The assumption of constant area proved inaccurate due to shearing of the soil.*

#4: Soils Cadets Were Consultants to Capstone Cadets

The fourth component of the integration was that the cadets in the soils class acted as consultants to the capstone cadets on their design project. A second individual study (CE489) was also completed by one cadet where he developed a Visual Basic (Microsoft, 1995) software package for footing design called: Footing Quick Solver (Conway, 1999, unpublished). This software package was ultimately used by at least one capstone group as part of their foundation design effort. Figure 5 shows foundation computations and screen captures from the Footing Quick Solver Visual Basic software package. This screen capture shown only a generic footing design computation, not one used for the project. The Footing Quick Solver was programmed to be used for:

- Field or laboratory soil property data
- Square, round, or rectangular spread footings
- Sandy or clayey soil

- Terzaghi's general bearing capacity formula
- Immediate settlement only

Capstone students were responsible for the footing design, but consultation with students in soils or with design aids like the Footing Quick Solver were highly encouraged. Consultation between students in soils and the capstone students also served to improve the CE371 cadets' understanding of the soil properties and how they affect the structure being designed in the Capstone. As previously described, students treated the test data as real data to be used for preliminary project design. Thus, it was apparent that they were putting slightly more care and consideration into the laboratory testing, discussion of results, and subsequent consultation.

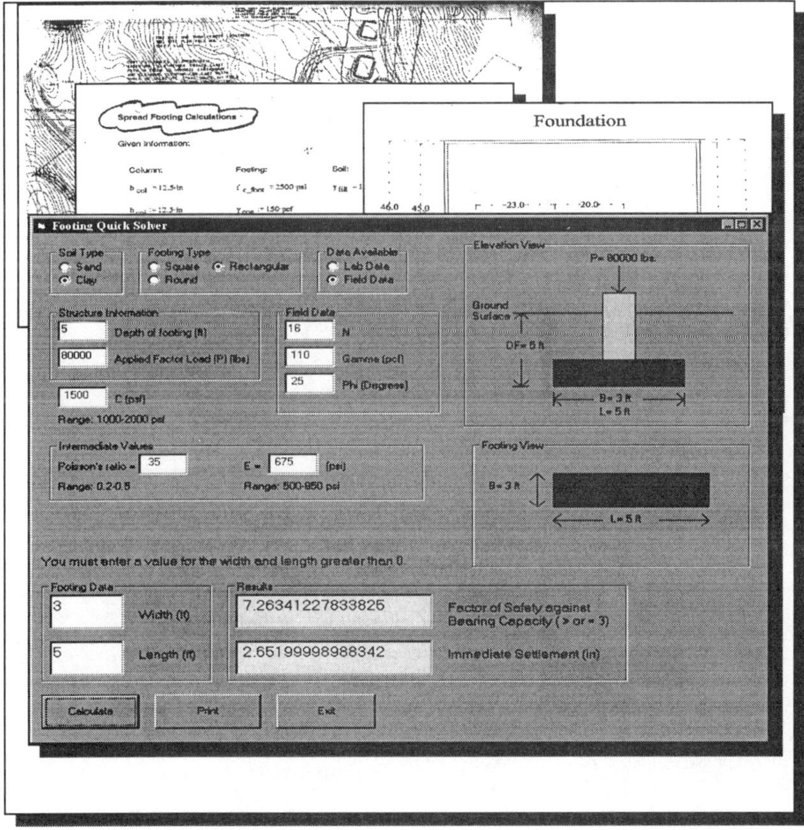

Figure 5. Foundation Design Calculations and Footing Quick Solver Visual Basic Software Package.

#5: Professional Brief to Clients

The fifth and final component of the integration was the professional brief that both the senior capstone cadets and their consultants, the junior-level soils cadets, gave to the Association of Graduates and to potential project donors. The capstone cadets prepared professional design reports including architectural and structural renderings and models as shown in Figures 6 and 7. The Soils cadets also prepared summary reports and display boards showing the soil stratigraphy, types, and properties, along with the design aids they prepared.

Cadet Assessment

Cadets were asked to provide anonymous comments regarding various aspects of the Soils course. Representative comments are included below not necessarily as an assessment of our various initiatives, but to provide anecdotal evidence of what was positively received and what was not. Only cadet comments related to the initiative to provide an integrated laboratory and classroom experience are included here.

Identify strengths of this course:

- *Incorporation of labs into daily material.*
- *Labs work well with [classroom] learning.*
- *Tested real dirt for a purpose.*
- *Revolving the work around the golf facility was great.*
- *Great course, primarily because it was always easy to see how the material we were learning was applicable to the real world.*
- *I thought the ongoing lab really reinforced concepts and gave a tie to real world applications.*
- *Tying the labs to the golf building design really reinforced real world applicability of soils.*
- *Labs were the best because it focused on a real world problem and how engineers test and solve for design problems*
- *One of the best practical classes I have taken.*
- *I liked going out in the field and testing soil that we collected.*
- *The usefulness of material in real world.*
- *Yes very applicable to the real world. I feel like I have contributed to the USMA Golf Facility.*

Identify weaknesses of this course:

- *The [2-hour] lab hour classes were killers.*
- *I hated the lab hours when we just had a 2 hour class.*
- *Long lab classes.*
- *2-hour in class labs.*

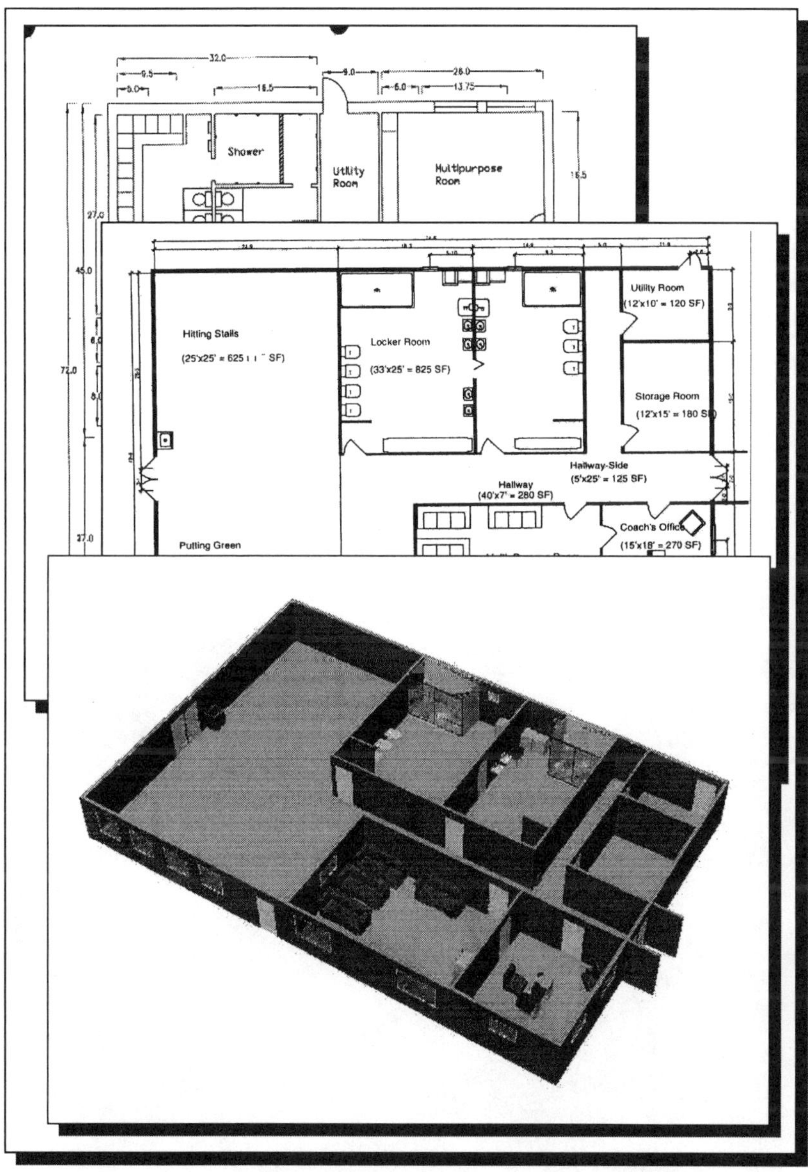

Figure 6. Architectural Design Plans and Renderings.

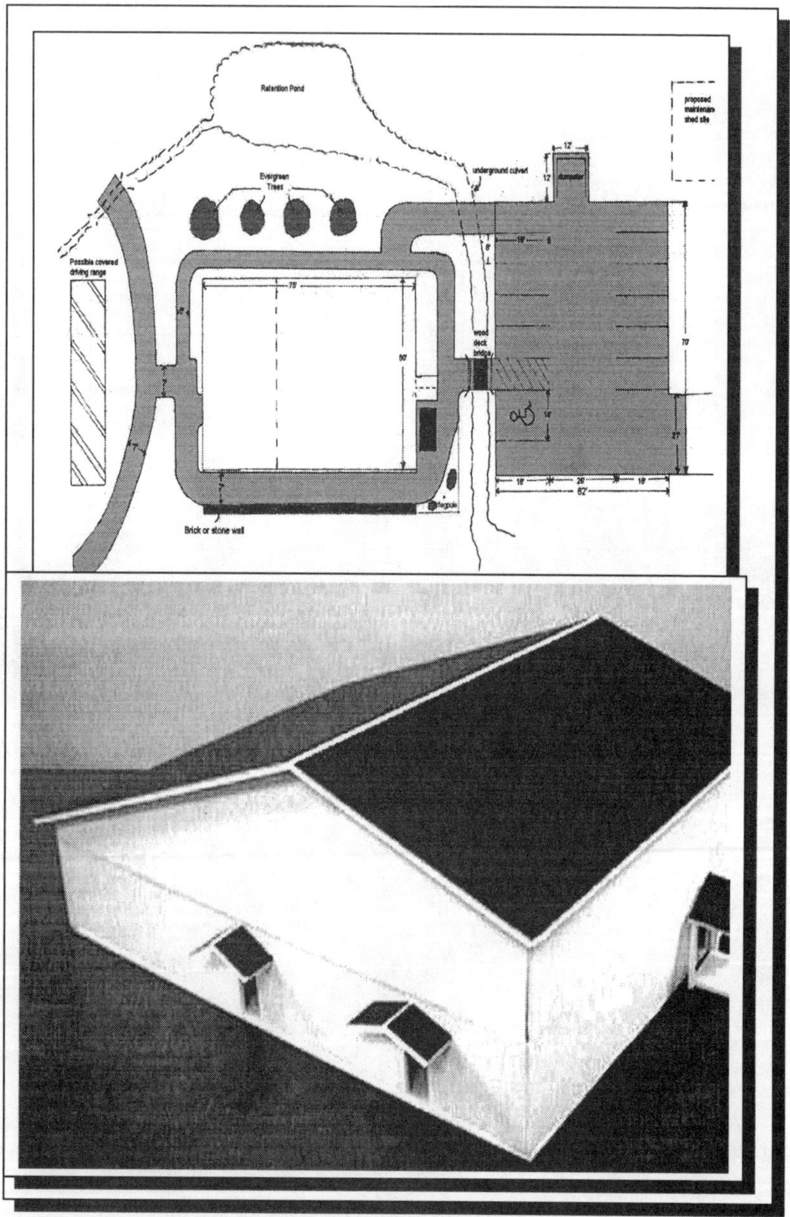

Figure 7. Site Plan Layout and Architectural Model for Client Briefing

It appears that the cadets liked and appreciated the integrated course, lab, and golf facility project. The real (and close to home) nature of the project helped them to see and appreciate how the laboratory testing and soil properties tied into the project design. The only pervasive negative comment was that cadets did not appreciate the long laboratory/ classroom sessions and two-hour lectures resulting from the interwoven laboratory and classroom hours.

Concluding Comments

This effort was a great success on all fronts -- the CE cadets and faculty got a great learning experience; the Cadet Golf Team got a jump start on a badly needed facility, and the USMA Alumni group got a free concept design that they could use to market the project. The only downside to such an integration as described here is the considerable care, coordination, and monitoring required of all the faculty. Additional monitoring is especially required of a project with considerable visibility; this one was highly visible outside the CE department, with some visibility even outside the Academy.

Shortly after the cadet briefs, the Alumni group received a generous donation of $500,000 that should fund a large portion of the new project design and construction. The Cadet Golf Team Practice Facility is now proceeding toward final design by a professional architect-engineer, with construction planned in about a year.

Acknowledgments

MAJ Richard Shelton, MAJ Brian Green, and MAJ Steve Hart initiated this cadet project in the CE division at the USMA. They were also the faculty involved in the initial Individual Study and the Soils and Capstone courses.

References

Albano, L. and Salazar, G. (1998). "Project-Based Course for Integration of Design and Construction at WPI", Journal of Professional Issues in Engineering Education and Practice, ASCE, vol. 124, no. 4, pp 97-104.

Allyn, Benshoof, Fagerquist, Jones, Lopez, Marshall, and Zadnichek (1999 unpublished). "Capstone Design Report Briefing", CE492, CE Division, Department of Civil and Mechanical Engineering, USMA, West Point, Spring Term.

Anderson, D. (1992). "Civil Engineering Capstone Design Course", Journal of Professional Issues in Engineering Education and Practice, ASCE, vol. 118, no. 3, pp 279-283.

ASTM (1999). "Annual Book of Standards", 100 Barr Harbor Drive, West Conshohocken, PA.

Conway, G. (1999, unpublished). "Using Visual Basic 4.0 for Footing Design", Final Report, CE489, Advanced Individual Study, CE Division, Department of Civil and Mechanical Engineering, USMA, West Point, Spring Term.

Evans, M. (1995). "Student and Faculty Guide to Improved Technical Writing", Journal of Professional Issues in Engineering Education and Practice, ASCE, vol. 121, no. 2, pp 114-122.

Fiegel, G. and DeNatale, J. (1998). "Hands on Geotechnical Engineering at the Undergraduate Level", Proceedings of the 1998 ASEE Annual Conference, Seattle, WA, June.

Gokhale, S. and Aldrich, J. (1997). "Integrating Community Service in the Construction Technology Curriculum", Proceedings of the American Society for Engineering Education, Milwaukee, WI, June 15-18.

Jones, S. (2000). "Using Senior Design Capstones as Model for Graduate Education", Journal of Professional Issues in Engineering Education and Practice, ASCE, vol. 126, no. 2, pp 83-88.

Lamancusa, J. Soyster, A., and George, R. (1997). "Industry-Based Projects in Academia – What Works and What Doesn't", Proceedings of the American Society for Engineering Education, Milwaukee, WI, June 15-18.

McEwen, E. (1994). "Integrated Capstone Design Experience", Journal of Professional Issues in Engineering Education and Practice, ASCE, vol. 120, no. 2, pp 212-220.

Melin, N. and Workman, R. (1999, unpublished). "Soil Analysis Report", CE371, Soils and Foundations, CE Division, Department of Civil and Mechanical Engineering, USMA, West Point, Spring Term.

Meyer, M. and Jacobs, L (2000). "A Civil Engineering Curriculum for the Future: The Georgia Tech Case", Journal of Professional Issues in Engineering Education and Practice, ASCE, vol. 126, no. 2, pp 74-78.

Microsoft Corporation (1995). Microsoft Visual Basic for Windows Version 4.0, Seattle WA.

Morris, C. and LaBoube, R. (1995). "Teaching Civil Engineering Design: Observations and Experiences", Journal of Professional Issues in Engineering Education and Practice, ASCE, vol. 121, no. 1, pp 47-53.

Pauschke, J. and Ingraffea, A. (1996). "Recent Innovations in Undergraduate Civil Engineering Curriculums", Journal of Professional Issues in Engineering Education and Practice, ASCE, vol. 122, no. 3, pp 123-133.

Sender, M. (1998). "Design of the Learning Environment: Professional-Project-Based Learning in Construction Education", Proceedings of the American Society for Engineering Education, Seattle, WA, June 28 – July 1.

USMA (1991). "Standards for Technical Reports", Department of Civil and Mechanical Engineering, USMA, West Point, NY.

GEOTECHNICAL ENGINEERING FOR ELEMENTARY SCHOOL STUDENTS

By

Gregg L. Fiegel,[1] Associate Member, ASCE, Victor Elia,[2] Student Member, ASCE, and Morgan Griffith,[2] Student Member, ASCE

ABSTRACT: Discussed is a geotechnical engineering demonstration designed to educate elementary school students about the geotechnical and civil engineering professions. The demonstration includes in-class experiments where students learn to classify different soils according to their grain size, permeability, and plasticity. The experiments are hands-on and are closely monitored by undergraduate student assistants. A second part of the demonstration allows students to observe a drilling operation. Cal Poly's drill rig is used to excavate a shallow bore hole and to retrieve an intact soil sample. Students are asked to classify the soil sample based on what they have learned in the classroom. To date, geotechnical engineering demonstrations have been completed at three elementary schools. Summary evaluations indicate that these demonstrations have entertained the students and sparked interest in engineering. The results of these evaluations are briefly summarized in this paper. Also summarized is the process that the authors followed in developing the demonstrations. It is hoped that the information presented will be valuable to geotechnical engineering educators and/or geotechnical engineering practitioners planning similar demonstrations in the future.

INTRODUCTION

One of the primary goals of the Cal Poly, San Luis Obispo ASCE Student Chapter is to educate local K-12 students about the profession of civil engineering. To accomplish this goal, members of the Student Chapter frequently organize simple engineering exercises and demonstrations for elementary, junior high, and high school age students in the San Luis Obispo area.

[1] Associate Professor, Department of Civil and Environmental Engineering, California Polytechnic State University, San Luis Obispo, CA 93407.

[2] Undergraduate Student Assistant, Department of Civil and Environmental Engineering, California Polytechnic State University, San Luis Obispo, CA 93407.

Recently, the authors, with help from the ASCE Student Chapter, developed a geotechnical engineering demonstration for elementary school students. The 1½ hour long demonstration is divided into four parts. First, through an informal discussion, the elementary school students are introduced to terms important in civil engineering and geotechnical engineering. Then, while still in the classroom, the students are divided into groups of 6 to 10 students each. In their groups, the students perform several geotechnical engineering experiments. The experiments are designed to illustrate such concepts as permeability, grain size, plasticity, water content, and soil classification. All of the experiments are hands-on and are closely monitored by undergraduate student instructors.

After the in-class experiments are completed, the elementary school students are shown a brief slide show on geotechnical engineering. Several slides illustrate how a drill rig is used to retrieve soil samples for laboratory testing. After the slide show, the students are led outside where they each receive a hard-hat and safety instructions. The students then watch as Cal Poly's drill rig is used to excavate a shallow bore hole and to retrieve an intact soil specimen. The students are asked to classify the soil sample and develop a simplified boring log based on their visual examination of the soil. Finally, the demonstration ends with a brief summary as well as a question and answer session involving the undergraduate student instructors.

To date, geotechnical engineering demonstrations have been completed at three elementary schools. Summary evaluations prepared by the elementary school students, the elementary school teachers, and the undergraduate student instructors indicate that these demonstrations have been very successful. The results of these evaluations are briefly summarized in the paper. Also summarized is the process that the authors followed in developing the demonstrations. It is hoped that the information presented will be valuable to geotechnical engineering educators and/or geotechnical engineering practitioners planning similar demonstrations.

TEACHING METHODOLOGY

The first step of the project was to research methods for effectively teaching elementary school students. This proved to be an interesting challenge given our limited experience in the area of elementary school education. The focus of the research was narrowed by targeting the demonstrations for students in the 4th, 5th, and 6th grades. Information was gathered from the education literature, the internet, and from interviews with professionals practicing in the field of education. Described below are the guidelines that were followed in developing the demonstration.

Small Group Instruction

Small student groups including up to ten students were used during most of the demonstration. Many of the elementary school instructors that were interviewed as part of this study recommended the use of small groups when presenting the material. The instructors, as well as education researchers (e.g. Phillips 1973), have recognized that small group demonstrations maximize student participation. The California Department of Education has recognized that exercises conducted in small, diverse groups help to facilitate cooperative learning (Holt 1993).

It should be noted that there can be some problems associated with organized group demonstrations. For example, vocal, outgoing students will tend to dominate the activities of the group while the more timid students will tend to observe rather than participate (Hare 1976). One way to avoid such a problem is to include hands-on experiments that require the participation of every student. A second way is to limit the number of students in each demonstration group. Both solutions were incorporated into the demonstration used in the project: hands-on work was the norm, and student groups were limited to ten members, as recommended by Shaw (1971).

During our interviews, one elementary school instructor stressed that questions asked of students in a group should be posed in such a way that all students are encouraged to formulate their own answer. Therefore, during the demonstration, students were asked a question and given time to formulate an answer. A raised hand was an indication of an answer, and a student was called upon only after everyone had raised his or her hand. This procedure further discouraged the influence of more vocal group members (Hare 1976; Schultz 1989).

Student Interest

A major concern in developing the geotechnical engineering demonstration was length. In other words, for how long would a 4th, 5th, or 6th grader be interested in the material being presented? Partin (1995) and other researchers recommend ten to fifteen minutes as a good rule of thumb for presentations focusing on one particular subject. In general, instructors interviewed as part of this study recommended a similar time frame, with a limit of twenty minutes for demonstrations involving hands-on activity. As a result, it was decided to limit all demonstrations and discussions focusing on a single subject to fifteen minutes.

A typical geotechnical engineering demonstration included an "icebreaker" and a "hook;" two brief activities that were designed to calm the students down and to generate initial student interest. These activities were followed by more in-depth discussions and exercises. During these discussions and exercises, visual aids and hands-on activities were employed as often as possible. Elementary school instructors and engineering educators alike have recognized the benefits of hands-on activities in education (e.g. ASEE 1986; Partin 1995; Wankat and Oreovicz 1993). In general, students show more interest for longer periods of time for activities incorporating hands-on work when compared with students subject to the more traditional auditory (lecture-based) mode of instruction (Partin 1995).

Finally, an attempt was made to incorporate mathematics, language arts, and earth science concepts into the demonstration. For example, students were asked to calculate water content and to develop a field boring log. A successful and interesting elementary school teaching lesson is one that incorporates multiple subjects that are already familiar to the students (Partin 1995).

Student Discipline

Occasionally, students of this age group have short attention spans and disciplinary problems. However, a formal discipline system was not instituted during the geotechnical engineering demonstration for a number of reasons. First, the elementary school teacher was required to be present during the entire demonstration. His or her discipline system could then be enforced thereby avoiding conflicts between the instructor and the guest lecturers. Second, it was felt that the negative connotations associated with discipline would detract from the objective of the project, which was to generate interest in the field of civil engineering. Finally, the hope was that an organized and exciting presentation would encourage the students to participate in a positive manner (Gray 1969).

EDUCATIONAL OBJECTIVES

Educational objectives were formulated prior to the formal development of the elementary school demonstration. It was decided that the primary objective of the demonstration would be to introduce students to the civil and geotechnical engineering professions using a hands-on, entertaining approach. Secondary objectives included: (1) to introduce basic soil classification procedures and techniques; (2) to illustrate the importance of proper soil classification in geotechnical engineering; (3) to incorporate basic math, language arts, and earth science exercises; (4) to encourage cooperative learning among team members; (5) to illustrate the scientific method; and (6) to emphasize the importance of a college education. The primary and secondary objectives were used as a guide when deciding on both the content and format of the geotechnical demonstration.

DESCRIPTION OF THE DEMONSTRATION

The recommended and researched teaching methods were combined with the educational objectives to develop the geotechnical engineering demonstration. A general outline of the demonstration is included as Table 1. The table lists the six main components of the demonstration: Introduction, In-Class Experiments, Slide Show, Drill Rig Activity, Conclusion, and Evaluation. The objectives of each demonstration component are briefly described in the table. In addition, the approximate time duration for each demonstration component is provided.

Table 1. Outline of the Geotechnical Engineering Demonstration

Demonstration Component	Primary Objective(s)	Duration (min)
Introduction	Grab the students' attention and stimulate interest in the subject matter	5
In-Class Experiments	Provide hands-on activities related to geotechnical engineering and soil classification	45-60
Slideshow	Illustrate geotechnical engineering projects and describe field investigations	10
Drill Rig Activity	Demonstrate a geotechnical field investigation (drilling and sampling)	15
Conclusion	Demonstrate the importance of soil classification in geotechnical engineering	5
Evaluation	Understand what went well and what did not during the demonstration.	5

Introduction

The introduction to the presentation included an "icebreaker" which required the students, as a group, to complete a short quiz. The quiz included basic questions related to classroom safety and acceptable behavior (e.g. "When talking in small groups, the best way to get your leader's attention is to..."). The objectives of the icebreaker were to allow the students to settle down, to remind the students to always think about safety, and to encourage the students to feel comfortable both asking and answering questions.

Following the icebreaker a "hook" was used to pique student interest. In this study, the hook was designed as a question and answer session focusing on the subject of civil engineering. The undergraduate student instructors described how common civil engineering works impact the everyday lives of everyone in the community. Numerous civil engineering examples were used, though roads and bridges seemed to be the most illustrative. A water supply example always seems to open the students' minds and allow them to think about civil engineering in a different way.

Eventually, the introductory component of the demonstration focused on geotechnical engineering. In California public schools, students in the 4th grade have already been introduced to basic geologic concepts. Therefore, the geotechnical portion of the presentation began with a familiar visual: students were given the opportunity to study a small-scale geologic cross-section that was constructed within an aquarium. The cross-section included different layers of soil and rock as well as a groundwater table. A cross-section of the earth was chosen as a prop for the introduction because it was a familiar visual (Partin 1995). Most students, therefore, were able to quickly relate to the demonstration.

In-Class Experiments

The elementary school students were asked to perform several simple soil classification experiments following the introductory presentation. A total of four experiments were performed: (1) Grain Size Distribution; (2) Water Content; (3) Permeability; and (4) Plasticity. These experiments, which are described in more detail below, were performed within the classroom under the direct supervision of two undergraduate student instructors. A mobile geotechnical laboratory was brought to the elementary school to facilitate the in-class experiments.

The experiments were completed by the students following a rotation scheme over a 45-60 minute period. The class was divided into four groups of approximately ten students each. Each of these groups tackled a single experiment for a period of about fifteen minutes. Enough testing equipment was brought to the school to allow the ten person groups to be subdivided into three groups of three or four students each. Thus, all of the experiments were completed using student groups composed of a maximum of four students.

Experiment 1: Grain Size Distribution

The objective of the grain size distribution experiment was to illustrate that different soils contain different size soil particles. To illustrate this concept, a simple sieve analysis was completed (see Figure 1). The experiment began with the instructor describing the important grain sizes and how they are related to soil classification. Gravel was defined as a soil retained on the #4 sieve while fine grained soils (clay and silt) were defined as soils passing through the #200 sieve. Sands were defined as soils with particle sizes falling in between the #4 and #200 sieves.

Figure 1. Students Perform a Sieve Analysis

After the initial discussion, the three student sub-groups were given a pre-prepared soil sample: one a gravel, one a sand, and one a clay. The students were then asked to visually inspect their soil samples and hypothesize as to the soil type. Next, the students were given a #4 sieve, a #200 sieve, a pan, and a lid. They were then asked to carefully sieve their soils through the sieve stack. Each student was given a turn shaking the stack. After shaking was complete, the students were asked to classify the soil based on what was observed on the sieves. Each sub-group was then asked to discuss and justify the conclusions that were reached.

Experiment 2: Water Content

The purpose of this experiment was to show students how to calculate a water content value. During this experiment the instructors discussed the importance of mathematics in engineering and explained how different soils may have different water contents. Water content, it was explained, can influence a soil's behavior.

Student sub-groups were each given three plastic sample jars. One jar contained a moist soil sample, one jar contained an oven-dried soil sample, and one jar contained water. Sand, gravel, and clay soils were used. The instructor explained that the moist and dry soil samples represented equivalent "amounts" of soil and that the water in the moist sample was represented by the water included in the third jar. Weights of the soil and water samples were noted on the jars using permanent marker. Students were asked to calculate the soil's water content based on the weights provided. A worksheet was supplied along with a calculator to simplify the computations. After determining the water content, the students were asked to assign a moist, dry, or wet classification using general water content ranges as a guide.

Experiment 3: Permeability

The objective of the permeability experiment was to illustrate the fact that water flows at a different rate through gravel, sand, and clay samples. The experiment began with the instructor defining permeability as a soil property which describes the rate at which water flows through a soil. Furthermore, gravel was defined as a soil with a very high permeability, sand was defined as a soil with a medium permeability, and clay was defined as a soil with a very low permeability.

The instructor next described the testing apparatus. The apparatus consisted of three large plastic jars, each with a six-inch high, two-inch diameter PVC column attached. The bottom of each PVC column was fitted with a porous cloth. Each column was half filled with one of the three soil types listed above.

The experiment was performed again using three sub-groups. One representative of each group was allowed to pour water into one of the PVC columns. An equal volume of water was poured by each representative simultaneously so that relative permeabilities could be observed. Based on the relative permeabilities observed, each sub-group was asked to classify its soil type. Answers were then discussed.

Experiment 4: Plasticity

The purpose of the plasticity experiment was to illustrate that different soils have different plasticity characteristics. The experiment began with the instructor defining plasticity as the "stickiness" of a soil. Clay was defined as a soil that, when moistened, can be "formed into a ball and then rolled into a worm." Sand was defined as a soil that could be formed into ball, but not rolled. Gravel was defined as a soil that could not be formed or rolled.

The instructor then divided the students into three sub-groups. Each sub-group was given a plastic dish containing a moist sample of clay, sand, or gravel. Students were then asked to work with the soil using a plastic mat and some water. Through experimentation, the students classified the soils given the above definitions of soil plasticity. After each sample was classified, the instructor provided all of the student groups with moist clay to play with prior to rotating to the next experiment.

Slide Show

Once the in-class experiments were completed, the entire class was brought together for a brief slide show. In general, the change in presentation technique captured the students' attention. The vocal description of the slides was kept light and up-beat in order to hold student interest. The purpose of the slideshow was to describe specific aspects of geotechnical engineering. Different examples were illustrated using photographs from local projects. Several slides showed how, during a typical site investigation, soil samples are retrieved using a drill rig.

Drill Rig Activity

The fourth component of the geotechnical engineering demonstration utilized Cal Poly's truck-mounted drill rig (Fiegel and DeNatale 2000; Fiegel 1998). After the slide show, the students were led outside to one of their athletic fields where the drill rig was located. Each student was given a hard hat and safety instructions. The hard hats were not really necessary as the students were kept a safe distance from the drill rig during its operation. However, the hats created excitement and instilled in the students the importance of job safety (see Figure 2).

As part of the demonstration the students watched the drill rig in operation for several minutes as a bore hole was advanced five feet into the soil. At a depth of five feet, a standard split-spoon sampler was driven into the ground to retrieve a soil sample. The students were allowed to examine the soil sample after it was brought to the surface and removed from the sampler. Each student was asked to record his or her observations by drawing a picture of the soil, by writing a brief description of the soil, and by classifying the soil. All observations were recorded on pre-prepared boring logs provided by the undergraduate instructors.

Figure 2. Students and the Rig at Santa Margarita Elementary School

The drill rig demonstration was performed for several reasons: (1) to allow the students to work with the heavy equipment in an outdoor environment so as to hopefully stimulate more interest in the field of engineering; (2) to incorporate additional writing and drawing exercises into the demonstration; (3) to illustrate to the students another aspect of geotechnical engineering; and (4) to have the students apply the knowledge and skills they had acquired in the classroom.

Conclusion

After the drill rig demonstration the students were brought back to the classroom for one final discussion. The discussion, which was usually kept short, allowed the undergraduate instructors to illustrate the practical importance of properly classifying soils. With the presentation finished, the students were allowed to ask questions. Though civil and geotechnical engineering questions were encouraged, many students were interested in hearing about college and the experiences of the undergraduate instructors. In two of the three elementary school demonstrations that were performed, lengthy question and answer sessions involving college life were initiated.

Evaluation

Finally, each demonstration was concluded with class, teacher, and undergraduate instructor surveys. The surveys helped the instructional team to evaluate the effectiveness of the presentations, exercises, and teaching methods. The feedback proved to be very helpful in the development and refinement of the demonstration.

TIPS FOR ELEMENTARY SCHOOL DEMONSTRATIONS

Over the past several months, the authors have gained a tremendous amount of experience in dealing with elementary school students. Listed below are some general tips that, in the authors' opinion, should help lead to successful laboratory and field experiences for elementary school students. These tips are based directly on the results of recently completed demonstrations.

Develop a Formal Lesson Plan

Prior to entering an elementary school classroom, a formal lesson plan was developed around a specific set of learning objectives. These learning objectives were formulated in meetings between the author and several undergraduate student instructors. The lesson plans were then refined based on conversations with local teachers as well as the results of practice sessions performed for undergraduates posing as elementary school students. This careful, up-front planning was, in our minds, successful, given the fact that the demonstrations worked very well without modification for three different schools.

Evaluate Your Work

In elementary school instruction, as in undergraduate instruction, it is important that your teaching be evaluated. Evaluation helps in the refinement of the presented material so that the students are guaranteed a valuable and rewarding demonstration. Upon completion of the geotechnical engineering demonstrations, the elementary school students, their teachers, and the undergraduate student instructors were each asked to complete evaluation surveys. The instructor surveys were detailed and included questions that focused on subject matter, teaching approach, and organization. Responses to these surveys were, in general, very informative.

The student surveys were greatly simplified compared to those presented to the instructors. The students were asked to rate, on a scale of 'bad' to 'great', different aspects of the demonstration. In addition, the students were asked to describe two new things that they learned as a result of the demonstration. It is interesting to note that these surveys did not provide much valuable information (see Figure 3). For the most part, everything was 'great', and the written comments generally showed that the students were "put on the spot."

From this experience, it was learned that thank you letters provide much more valuable evaluation information. Several days after one demonstration, the elementary school instructor asked her students to prepare thank you letters for the undergraduate student instructors. These letters were very detailed and provided a great deal of insight into what worked during the demonstration and what did not.

For future demonstrations, the plan is to continue to distribute student evaluation forms. However, elementary school instructors will be encouraged to have the students write thank you letters or short essays, given the previous success. Certainly, longer comment forms could be completed by the students after a demonstration. However, it has been found that time is very limited given the fact that a demonstration is already 1½ hours long.

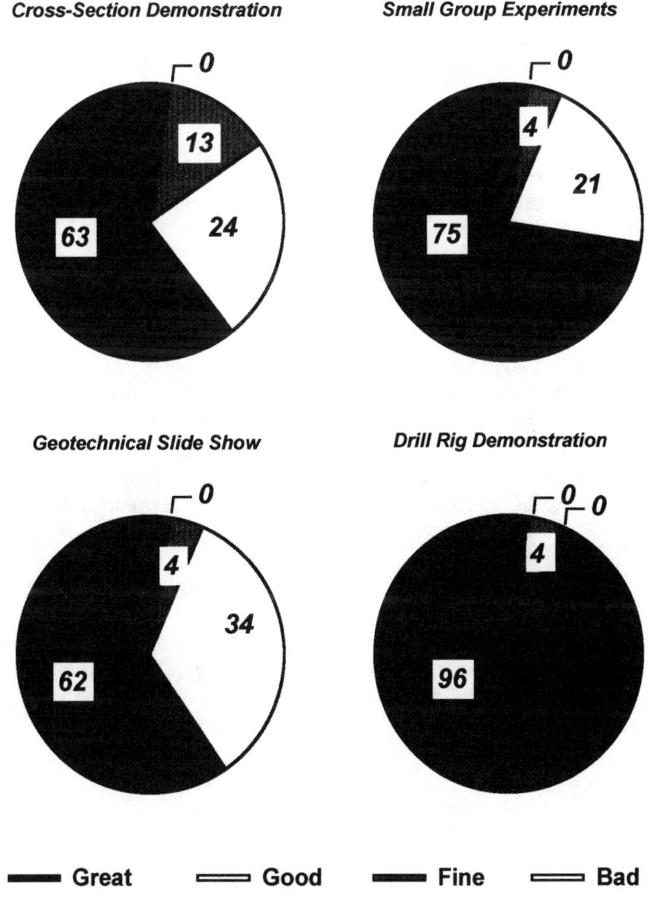

Figure 3. Student Responses to Evaluation Questions (100 Total). Students were asked to "Rate" Different Aspects of the Demonstration.

Recruit a Diverse Group of Instructors

It is important that the undergraduate student instructors relate to as many of the elementary school students as possible during a demonstration. Undoubtedly, the audience at an average elementary school level will represent an ethnically/gender diverse group of students. Therefore, its is recommended that the make-up of the instructors be as ethnically/gender diverse as the audience being addressed.

As an example, in the San Luis Obispo area, many of the classes are split between white male, female, and Hispanic males/females. In an attempt to "touch" all the members of the audience, female and Hispanic instructors were added to the team. This approach appeared to be very successful. In one class of 4th graders, an Hispanic instructor was able to speak Spanish with some students who had not yet mastered the English language. In another demonstration with 6th graders, a question and answer session on college life was carried-out after the geotechnical engineering demonstration had been completed. This session was led by a female instructor. Several of the female students were very interested in what engineering life is like for a woman.

Understand Your Audience

It is important that you research what can be expected when entering the classroom. You must understand your audience. For the authors, it was difficult to remember life as a 10-year old! Therefore, numerous questions were posed: What is the attention span of a student in elementary school? What math and science concepts are appropriate for elementary school students? What is the best way to deal with behavioral problems? Should the class and field demonstrations be completed in groups or with the entire class?

Fortunately, our colleagues in the education field have answers to all of the questions listed above. Initially, the above questions were answered by researching the appropriate educational literature. Further information was then gathered by interviewing faculty members of the Education Department on campus. Finally, a wealth of information was realized after meeting, in person, with the elementary school instructors. The valuable insight provided by these instructors is absolutely necessary for a successful presentation.

Practice, Practice, Practice

When entering an unknown environment, like an elementary school classroom, you should be as prepared as possible. Being prepared will help you to deal with unexpected problems. In addition, being prepared will allow your demonstration to have a more personal tone, which helps generate student interest.

Prior to conducting the geotechnical demonstrations, the Cal Poly undergraduate student instructors practiced in front of other undergraduates posing as elementary school students. In addition, prior to traveling to the schools, all undergraduate student instructors were given detailed written instructions and trained in the various experiments.

CONCLUSION

A recent drop in undergraduate student enrollment within the field of civil engineering has hinted that the promotion of engineering at the K-12 level may be necessary in the very near future. At Cal Poly, the principal author has begun working with the ASCE Student Chapter to develop a variety of civil and environmental engineering demonstrations/exercises for elementary, junior high, and high school students. A recently developed elementary school demonstration focuses on geotechnical engineering.

The purpose of this paper has been to describe the geotechnical engineering demonstration that the authors have developed. To date, three very successful demonstrations have been completed. Positive evaluations were received from the elementary school students and their teachers. These evaluations have allowed the authors to refine and further develop the demonstrations. Refinement has led to a comprehensive instructional guide which is available for Cal Poly civil engineering undergraduates interested in giving similar geotechnical demonstrations in the future.

The demonstration described in the paper can serve as a guide for engineers thinking of developing their own. Certainly, the components of our presentation are not essential. However, we do recommend that any engineering demonstration for elementary school children have a significant hands-on component. This appeared to be the most successful and well received aspect of the demonstration.

Finally, regardless of whether or not recent enrollment trends in civil engineering are a true concern or just an anomaly, the opportunity to work with the elementary school students and their teachers through engineering demonstrations is very enjoyable and extremely rewarding. The experience is highly recommended to geotechnical engineering faculty and practitioners alike.

ACKNOWLEDGMENTS

The demonstration described in this paper was developed, in part, with funding provided by the National Science Foundation's Division of Undergraduate Education through grant DUE #9751318. The authors gratefully acknowledge this support. The authors also acknowledge the financial support provided by the International Association of Foundation Drilling (ADSC) and numerous Cal Poly alumni who are currently engaged in the practice of geotechnical engineering.

Finally, the authors would like to thank the members of the Cal Poly ASCE Student Chapter. The hard work and dedication of these volunteers helped make the geotechnical engineering demonstrations very successful.

REFERENCES

American Society for Engineering Education, ASEE (1986), *Quality of Engineering Education Project: Final Report*, ASEE, New York, NY.

Fiegel, G.L. (1998). "Cal Poly Obtains a Truck-Mounted Field Exploration Rig." *Foundation Drilling Magazine*, International Association of Foundation Drilling (ADSC), Dallas, Texas, January, 27-31.

Fiegel, G.L. and DeNatale, J.S. (2000). "A Hands-On Undergraduate Geotechnical Engineering Program." *Proceedings, GeoDenver2000 Specialty Conference*, ASCE, Washington, D.C.

Gray, J. (1969). The Teacher's Survival Guide. David S. Lake Publishers, Belmont, California.

Hare, Paul A. (1976). Handbook of Small Group Research. The Free Press, New York, New York.

Holt, D. D. (1993) "Cooperative Learning for Students from Diverse Language Backgrounds: An Introduction." *Cooperative Learning: A Response to Linguistic and Cultural Diversity*. Delta Systems and Center for Applied Linguistics, 1-8.

Partin, R.L. (1995). Classroom Teacher's Survival Guide. The Center for Applied Research in Education. West Nyack, New York.

Phillips, G.M. (1973). Communication and the Small Group. Bobbs-Merrill Company, Inc., New York, New York.

Shaw, M. (1971). Group Dynamics: The Psychology of Small Group Behavior. The McGraw-Hill Book Company, New York, New York.

Schultz, B.G. (1989). Communicating in the Small Group. Harper & Row Publishers, New York, New York.

Wankat, P.C. and Oreovicz, F.S. (1993), Teaching Engineering, McGraw-Hill, New York, New York.

NUMERICAL AND VISUALIZATION TECHNIQUES IN GEOTECHNICAL ENGINEERING EDUCATION

Muniram Budhu[1]

Abstract

Numerical methods, especially the finite element method, are widely used in geotechnical engineering to analyze a large range of boundary value problems. The outputs from numerical methods are often overwhelming and one has to wade through a large body of numbers to get the desired results. Software technology is now available to animate the results of numerical methods to provide visual outputs. These visual outputs help to improve our interpretation and evaluation of numerical analyses. Significant progress in visualization has been made in commercially available programs such as ABAQUS. However, progress in the area of geotechnical education has been rather slow. The purpose of this paper is to present visualization techniques to help students understand fundamental concepts in geotechnical analysis. In this paper, a multimedia interactive and animated teaching module on the basic concepts of one-dimensional consolidation of soils using a finite difference scheme is presented. Key learning objectives are described and then self-paced interactions are developed to meet these objectives.

One-dimensional Consolidation Concepts

Conventional Approach

An essential concept in geotechnical engineering is the process of soil consolidation. The concept is conventionally explained in textbooks using a simple mechanical system consisting of a water container with a controlled opening, a spring and a load cap (Figure 1). The spring is intended to represent the soil solids and the water within the spring represents the pore water in a saturated soil medium. The control opening often has a pressure gauge to illustrate the excess

---

[1] Professor, Department of Civil Engineering & Engineering Mechanics, University of Arizona, Tucson, AZ, 85721

pore water pressure. A petcock on the load cap is used to represent the effects of permeability.

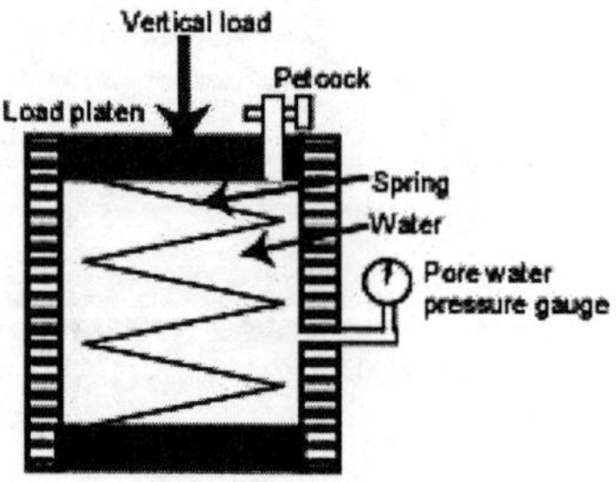

Figure 1 Conventional spring analogy for one-dimensional consolidation

There are two problems with this approach.

1. The simulation of the soil solids as an elastic spring is incorrect because, unlike the spring, the soil solids do not return to their original configuration upon unloading.

2. It does not provide information on how the initial excess pore water pressure dissipates with time and soil thickness. The students are told that at time $t = 0$, the initial excess pore water pressure is equal to the applied vertical stress and at time $t \Rightarrow \propto$ the initial excess pore water pressure dissipates. The students then have to await the presentation of Terzaghi's one-dimensional theory to learn how the initial excess pore water pressure dissipates with time.

Despite the two weaknesses cited above, the spring-water analogy has served to give, at least, some essential ideas on the consolidation process.

New Approach

Current technologies allow us to present a host of media element (multimedia) consisting of text, graphics, sound, animation and interactivity in a visual environment to enhance learning. Although many media elements can be combined to prepare a courseware, the developer has to use careful judgement to achieve interest and emphasis. Multimedia must connect the learner with the knowledge to be learned.

The purpose of the proposed approach is:

1. To give a coherent understanding of the process of consolidation.

2. To allow students to interact with the courseware to examine, in particular, "what-if" situations.

Numerical Solution for Visualization

The consolidation of soil is based on Terzaghi's one-dimensional consolidation equation

$$\frac{\partial u}{\partial t} = C_v \frac{\partial^2 u}{\partial z^2} \qquad (1)$$

that describes the variation of excess pore water pressure (u) with time (t) and depth z; C_v is the coefficient of consolidation which depends on the permeability of the soil and the modulus of volume change. The solution of this equation, like all differential equations, requires knowledge of the boundary conditions. For soils, the boundary conditions are complex and we cannot obtain closed form (analytical) solutions for many practical problems. We then resort to numerical methods. Three numerical schemes are popular - finite difference (FD), finite element (FE) and boundary element (BE).

An explicit time-marching FD scheme is used in this contribution because the solutions for equation (1) can be obtained quickly – no matrix needs to be stored and manipulated. The FD form of equation (1) is available in the literature. However, for ease of reference and completeness, the FD equation for one-dimensional consolidation is repeated here as

$$u_{i,j+1} = u_{i,j} + \frac{C_v \Delta t}{(\Delta z)^2}\left(u_{i-1,j} - 2u_{i,j} + u_{i+1,j}\right) \qquad (2)$$

where i, j refers to a nodal position, i denotes depth and j denotes time and Δ is a incremental quantity. Equation (2) is valid only for the nodes in the interior of the soil. At impermeable boundaries the FD equation is

$$u_{i,j+1} = u_{i,j} + \frac{C_v \Delta t}{(\Delta z)^2}\left(2u_{i-1,j} - 2u_{i,j}\right) \qquad (3)$$

For stability of the numerical solution the term $\frac{C_v \Delta t}{(\Delta z)^2}$ must be less than 0.5.

Visualization Scheme

The student is presented with an experimental arrangement consisting of a soil sample sandwiched between two porous stones in a cylinder (Figure 2). A load platen is placed on the top porous stone to load the soil. Water passing through the porous stones is channeled by tubes to a burette. A control value is located near the entrance of the burette to either allow water to flow in or not. A weight is used to apply a vertical stress to the soil sample. Animated pore water pressure transducers are positioned at the top, ¼ height and ½ height in the soil sample. Readout units are connected to the pore water pressure transducers to record and display the changes in excess pore water pressure with time. The student uses a mouse to drag the weight and place it on the load platen. The flow control valve is kept closed.

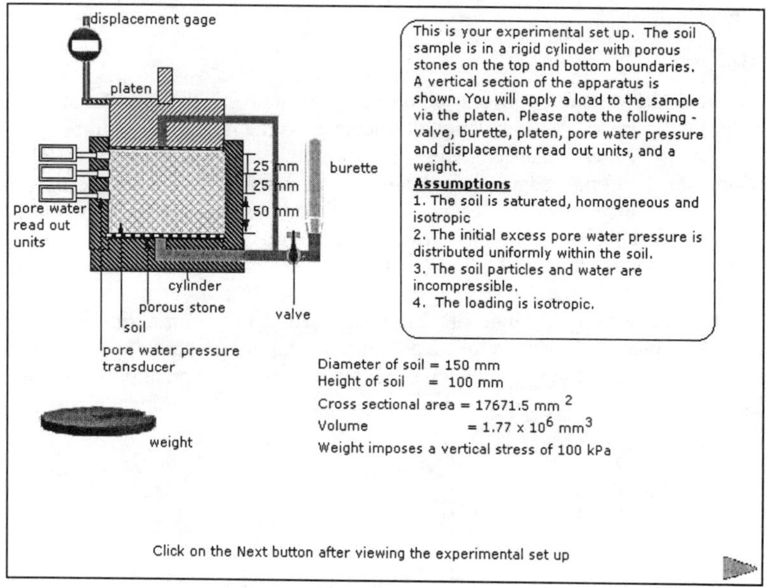

Figure 2 Model experimental set up

As soon as the load is on the load platen, a plot of the changes in excess pore water pressure with depth at time $t \cong 0$ is displayed. Two essential points are illustrated to the students (Figure 3). The first point is that when the load is applied to an isotropic soil without any water escaping from the soil, the initial excess pore water pressure is equal to the applied vertical stress. The second point is that no consolidation settlement occurs because water was prevented from draining out of the soil.

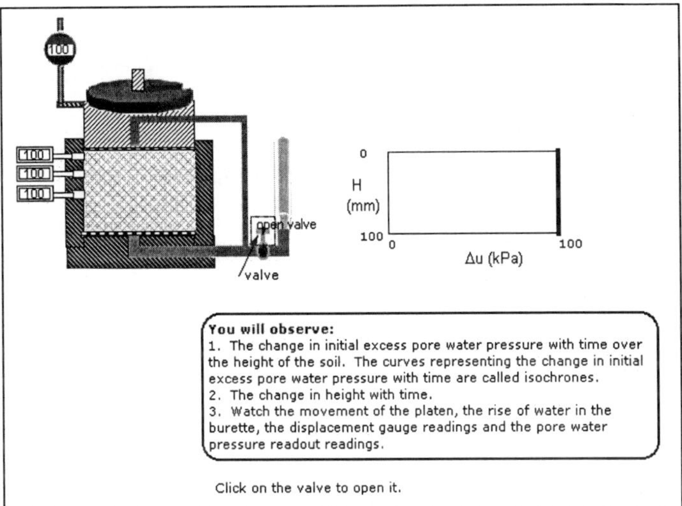

Figure 3 Excess pore water pressure as soon as load is applied.

In the next phase, the student uses a mouse to open the control value. The student then observes the following (Figure 4).

1. Water drains from the soil and is collected in the burette.
2. The rate of flow of water decreases with time.
3. The change in initial excess pore water pressure with time at different depths.
4. The settlement of the soil with time.

In this phase, the students learn the following.

1. The initial excess pore water pressure decreases with time.

2. The settlement of soil is non-linear – larger settlement occurs initially and then decreases slowly with time.

3. The settlement of the soil occurs as a result of the expulsion of water from the soil.

4. The amount of settlement is related to the amount of water expelled.

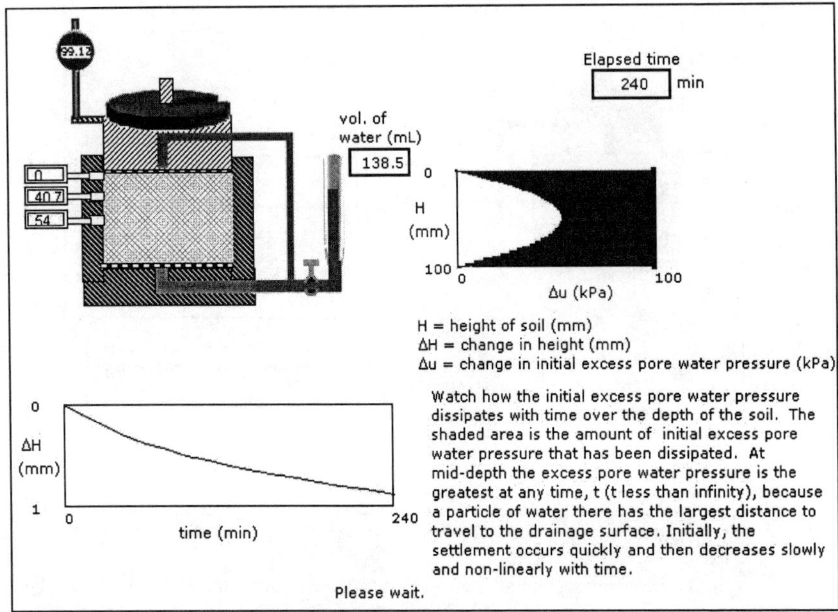

Figure 4 Spatial variation of excess pore water pressure with time and settlement with time

Exploration

The initial experimental set-up is for double drainage and the initial excess pore water pressure is assumed to be uniformly distributed. A student can now explore other situations. Three situations are given to the students (Figure 5). The first situation is single drainage with a linear distribution of initial excess pore water pressure. The second situation is again a single drainage situation but with a half-parabolic distribution of initial excess pore water pressure. The third situation is double drainage with a full parabolic distribution of initial excess pore water pressure. The student clicks on any one of these situations and observes the spatial distribution of initial excess pore water pressure with time. The maximum initial excess pore water pressure at a given time, e.g. 240 minutes, for each of these situations is summarized. The learning outcomes are:

1. The distribution of excess pore water pressure at a given time depends on the initial distribution of excess pore water pressure.
2. Double drainage causes more dissipation of initial excess pore water pressure than single drainage.

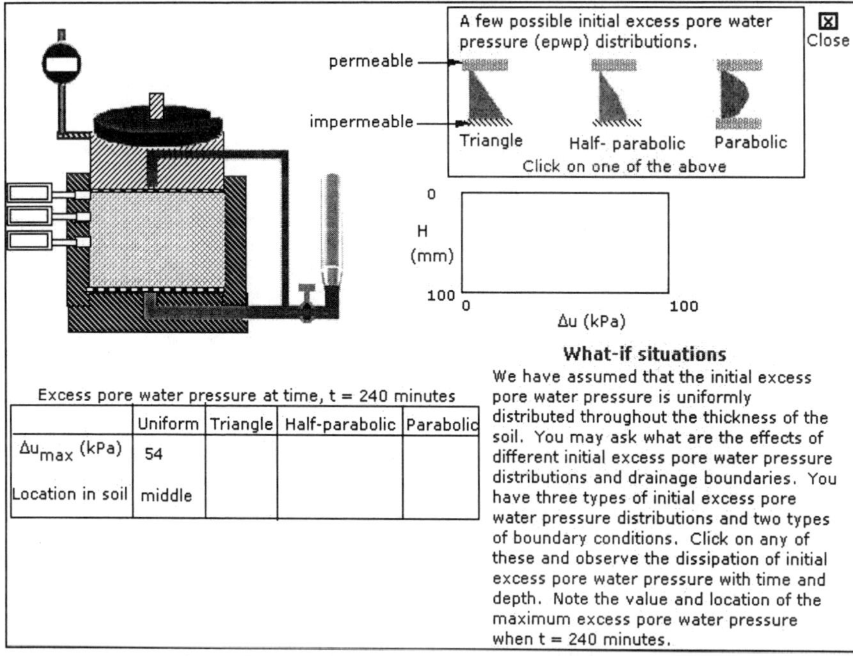

Figure 5 Examination of what-if situations

Effects of Loading

Rather than putting additional weights to continue loading the soil, a pneumatic cylinder is added to the apparatus whereby the student clicks on a control valve to open it and air pressure is used to load a ram sitting on the top of the load platen (Figure 6). For each increment of load the students add, the void ratio change is plotted on three graphs: (1) void ratio versus vertical effective stress (2) void ratio versus logarithm of vertical stress (3) volumetric strain versus vertical stress. After reaching a maximum load, the student is requested to unload the soil incrementally and then reload it beyond the previous maximum load. The void ratio change at each unloading increment is also plotted on the three graphs.

The student learns the following.

1. The form of the loading/unloading/reloading graphs on the three different plots using the same data.
2. That soil behavior is inelastic.
3. Once the previous maximum vertical stress is exceeded after unloading/reloading the soil continues along the path followed just prior to unloading.

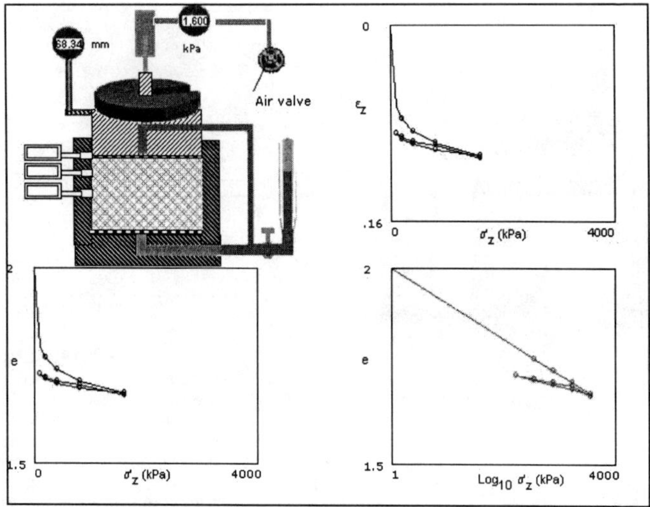

Figure 6 Loading, unloading and reloading interaction.

Interpretation of Consolidation Data

The next learning phase involves teaching the student how to interpret the data from the simulated test. They are taught how to determine compression and recompression indices, modulus of volume change and the meaning of preconsolidation stress and how to calculate overconsolidation ratio (Figure 7).

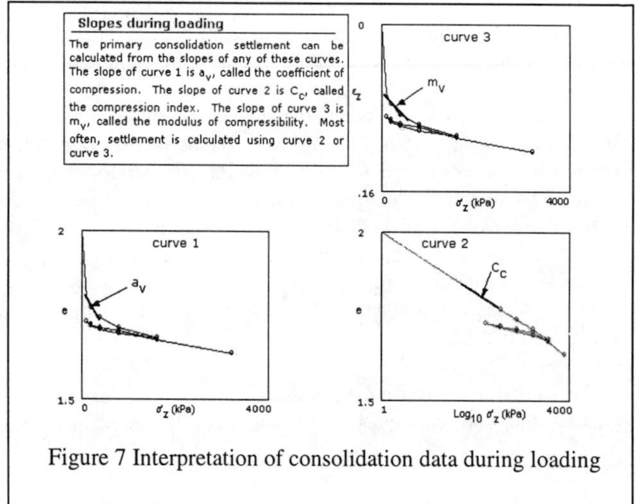

Figure 7 Interpretation of consolidation data during loading

Preliminary Evaluation

Thirty-six students in a soil mechanics class at the University of Arizona were given notes on the conventional spring analogy, notes and a CD-ROM on the methodology described here for teaching the basic concept of consolidation. One third of the class was told to read the notes on spring analogy, the notes on methodology described here, and then run the CD. The second third read the notes on the spring analogy and then ran the CD. The last third ran the CD first and then read the notes on the spring analogy followed by the notes on methodology described here. The intention of the above exercise was to determine if the CD-ROM (visual media) with its interactivity enhances learning and retention. When these students were asked which of the three gave them the best understanding of consolidation, ninety three percent indicated that the CD-ROM was the best. The other seven percent indicated that they have obtained enough information from reading the notes. A questionnaire was also administered. One of the statements in the questionnaire is "I am better able to visualize the ideas and concepts taught in this course." Half the students strongly agree while the other half agree with the statement.

Conclusions

In this paper a self-paced multimedia simulation of the essential concepts of one-dimensional consolidation was presented. The simulation was used to enhance learning and retention. Thc major conclusion reached from preliminary evaluation of the methodology presented is that multimedia simulation of the one-dimensional consolidation concept does enhance learning. More exhaustive evaluations involving several institutions are now been conducted.

Geotechnical Laboratory – A Multimedia Experience

Sunil Sharma, MASCE[1] and James H. Hardcastle, MASCE[2]

ABSTRACT

This paper discusses the development of ten modules for teaching undergraduate Civil Engineering students the experiential aspects of laboratory testing associated with the first course in geotechnical engineering. These modules may also be used in other disciplines with limited laboratory facilities and by technicians trying to get familiar with these types of tests. This courseware will range in complexity from visual soil classification techniques to triaxial shear strength testing. Each of these modules provides: (1) reference material, (2) modeling and simulation using rendered 3-D graphics and video, (3) an interactive tutorial and quiz to test the student's understanding, (4) handouts containing typical lab data, results and blank forms in Adobe PDF format, and (5) exploratory links to external World Wide Web sites. The series have been developed using ToolBook II from Asymetrix and 3D Studio Max from Kinetix.

In this paper, the Atterberg Limit Test module is presented to illustrate the overall framework of the ten modules. This module can show students the sequence of operations required to complete the liquid and plastic limit tests. These procedures, illustrated using video clips of the plastic limit test and an animation of the liquid limit test, provide the student with a realistic experience of the actual test *before* going to the laboratory.

Once in use, we expect that these modules will allow students to become more familiar with the laboratory equipment and procedures and should result in an improved appreciation for the uniqueness of soil and its behavior.

[1]Associate and [2]Professor, University of Idaho, Department of Civil Engineering Moscow, ID 83844-1022. ssharma@uidaho.edu, jimhard@uidaho.edu

INTRODUCTION

In the Department of Civil Engineering at the University of Idaho, multiple laboratory sections (classses) are used to ensure that students receive significant hands-on training that will enhance their educational experience and give them an accurate insight into the general practice of engineering. In order to provide such a rich experience, faculty expend considerable effort to ensure that the highest standards are achieved at all times.

To facilitate this learning experience, a Computer Based Learning (CBL) framework has been developed to enhance the laboratory instruction. It is also hoped that this will reduce the overall demands on the faculty time for multiple laboratory sections. These tools can operate on individual workstations or over a local area network (LAN). Future modifications will allow these modules to directly operate over a local Intranet or over the more global Internet for the convenience of off-campus students.

COMPUTER REQUIREMENTS

The modules, or courseware, have been designed to operate on a multimedia capable (i.e. one with a CD-ROM drive and sound card) IBM-PC, or compatible, with at least 16 Mb of RAM, running the Microsoft Windows-95, – 98 or Windows-NT operating system. Microsoft Internet Explorer (version 5, or later) must be installed on the system. Although it is not essential, it is recommended that the computer have access to the Internet for downloading small files, communicating with the instructor and accessing any web-links assigned by the instructor.

PROJECT SCOPE

The project has developed ten Computer Assisted Instruction (CAI) modules to supplement the existing geotechnical laboratory sessions. The 10 modules are:

1. Water content, weight-volume relationships
2. Visual-manual soil classification
3. Atterberg limits
4. Grain size distribution: both sieve and hydrometer analysis
5. Compaction tests for Moisture-Density relationships
6. Field density testing: soil unit weight, water contents, compaction control
7. Permeability tests: constant and variable head tests, fine and coarse grained soils
8. Consolidation testing

9. Direct shear test
10. Triaxial tests – UU, CU and CD

The courseware ranges in complexity from visual-manual soil classification techniques to triaxial shear strength testing with each module providing the following common features: (1) reference material, (2) modeling and simulation, (3) an interactive tutorial and quiz to test the student's understanding, (4) handouts containing typical lab data, results and blank forms in Adobe PDF format, and (5) exploratory links to external World Wide Web sites.

The modeling and simulation will build on the reference material and add animation effects to introduce the students to the actual sequence of the recommended laboratory procedures. The most important feature of the courseware will be the tutorial, which will provide interactivity through an elaborate "branching" structure that will use instructional games and drill and practice concepts to test the student's understanding of fundamental principles and test procedures before allowing them to proceed to the next topic.

MODULE FRAMEWORK

The overall framework for constructing a typical learning module is shown in Figure 1. The multimedia content, file types (i.e. AVI, WAV etc.) and the software used to generate or edit these items is shown in this figure. The items that can be modified are shown on the right side of this figure. For example, if a department has a different set of handouts, these can be either combined with the ones provided or, they can replace the ones provided with a new PDF file. Of course, the instructor will have to have access to Adobe Acrobat in order to generate the replacement file. Similarly, the on-screen display used to explain the calculations is a Rich Text Format (RTF) file, which can be readily generated using most word processors.

The most elegant changes that can be made involve the HTML help file that is provided with each module. These help files have a "hook" where an instructor can link his or her own local HTML files to broaden the subject, or even add links to a local intranet or the more extensive World Wide Web (WWW). With such links, the content of such a background information help-file becomes essentially limitless.

The modules have been prepared such that a student can complete the simulation within 25 – 35 minutes. It is generally recommended (... by educational psychologists!) that such exercises be limited to only thirty minutes as students usually lose interest if these are too long or overly complicated. One has to remember that the user will typically only complete a module once (or possibly twice) before proceeding to the next laboratory module. In view of these constraints, we have tried to prepare all the modules within a consistent and intuitive framework. To illustrate our approach, we will present the "Atterberg Limits" (AL) module as an example.

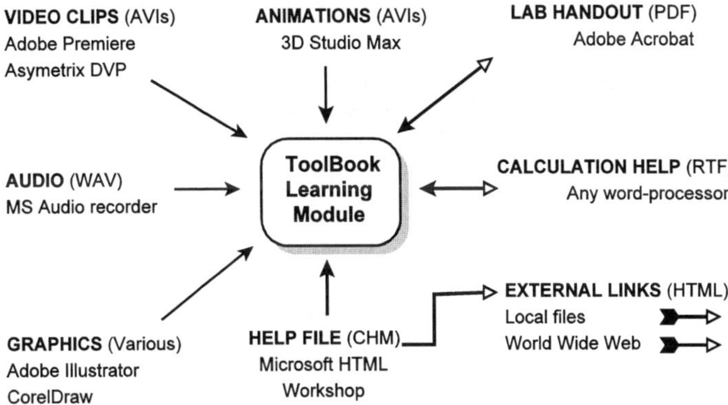

Figure 1. The common framework of multimedia content and potential extensibility.

ATTERBERG LIMITS – Example Module

Figure 2 presents the opening screen for the Atterberg Limits module. From this screen, the user can review the objective of the Atterberg Limits testing, which is just one short paragraph to allow for easy comprehension.

Selection of the "Background" option takes the user to a HTML-Help file which provides greater detail (see Figures 3 and 4) about the Atterberg Limits test. This is similar to the material that one would find in a text-book, but it can be supplemented with new material, video-clips and hyperlinks to other sections, a glossary and other WWW sites.

From the "Main Screen" (Figure 2), the "Liquid Limit Test" and the "Plastic Limit Test" selections take the user to the interactive material which illustrates the appropriate laboratory procedures for these tests.

As with all laboratory sessions, a set of brief notes, blank data forms and a set of example data (i.e. a *handout*) are available for printing. This printable material has been prepared in Adobe's "Portable Document Format" (PDF) as it offers the most convenient approach to pass on a high quality document to the student. A typical file is provided for use, but if the instructor wishes to change these notes, a new PDF file can be readily produced using Adobe Acrobat.

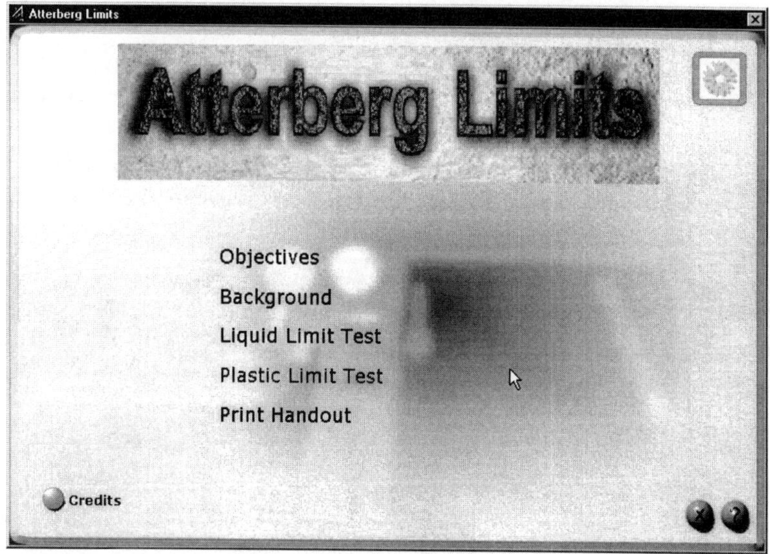

Figure 2. Main screen of the Atterberg Limits learning module

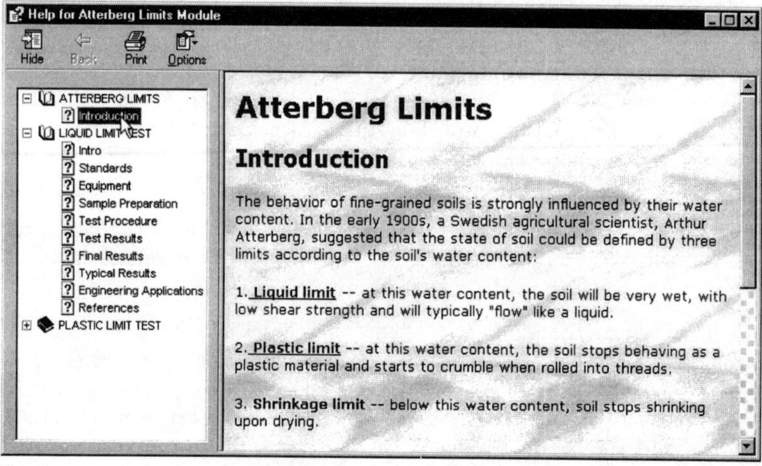

Figure 3. The HTML help file accessed from the "Background" option.

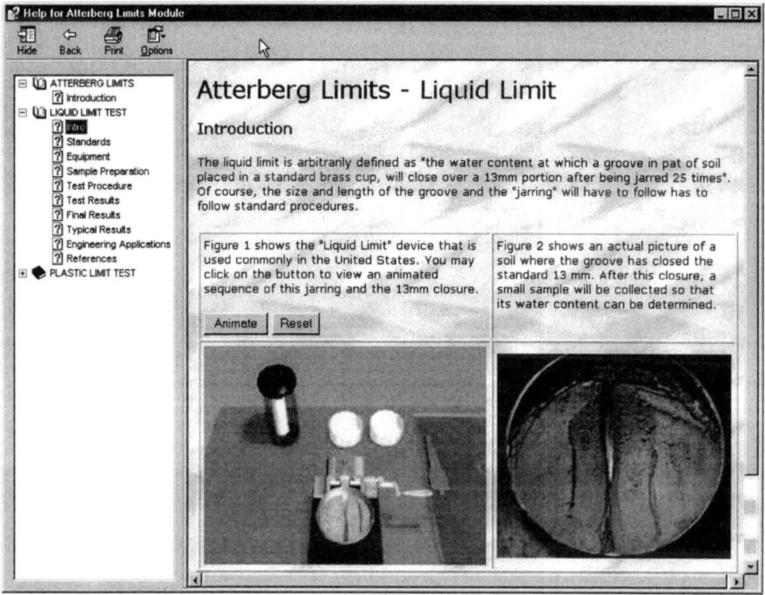

Figure 4. Link to thc information about the Liquid Limit test

Upon selection of the "Liquid Limit Test" option, the user will see the screen shown in Figure 5. This is the main part of the interactive module for the Liquid Limit test. From this screen, the user can press one of the *topical* buttons on the left to view a list of related items that are shown on the right. In this figure, the "Equipment" option has been selected and offers the student an opportunity to view the laboratory equipment that will be used for a typical liquid limit test.

By clicking on one of the list items (Liquid Limit Equipment), the student is presented with a view of the equipment that is commonly used for determining the liquid limits (Figure 6). The student can learn about the various pieces of equipment by moving the mouse-cursor over objects in the image. In Figure 6, for example, the "Casagrande Grooving Tool" is identified by the current location of the cursor. We find that if the student can become familiar with the terminology *before* attending the actual laboratory session, unnecessary questions and repetitive explanations can be avoided and more time can be dedicated to the actual test.

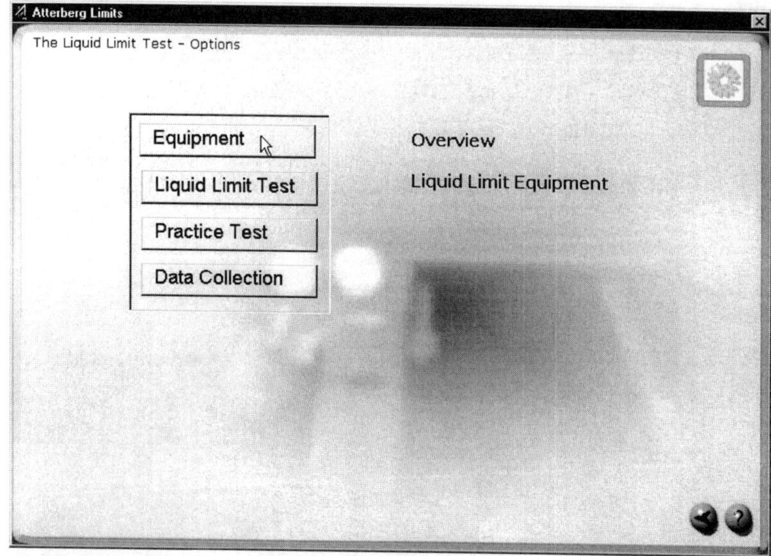

Figure 5. The Liquid Limits contents screen

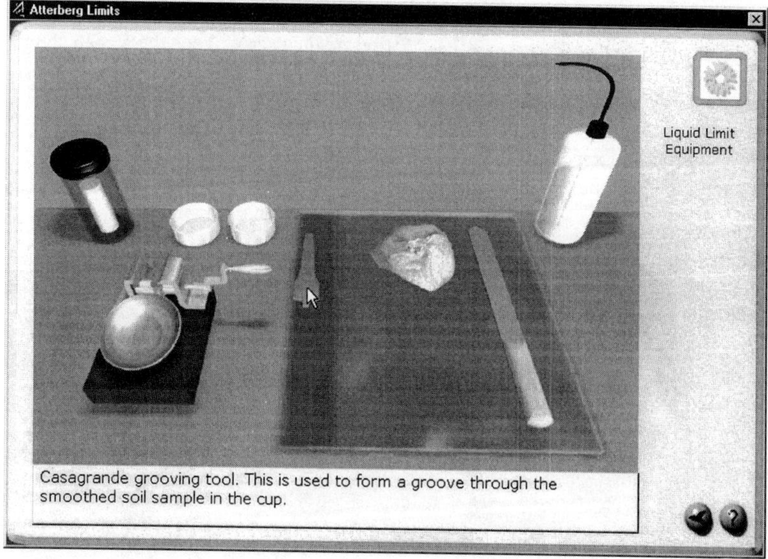

Figure 6. Equipment used for liquid limit testing

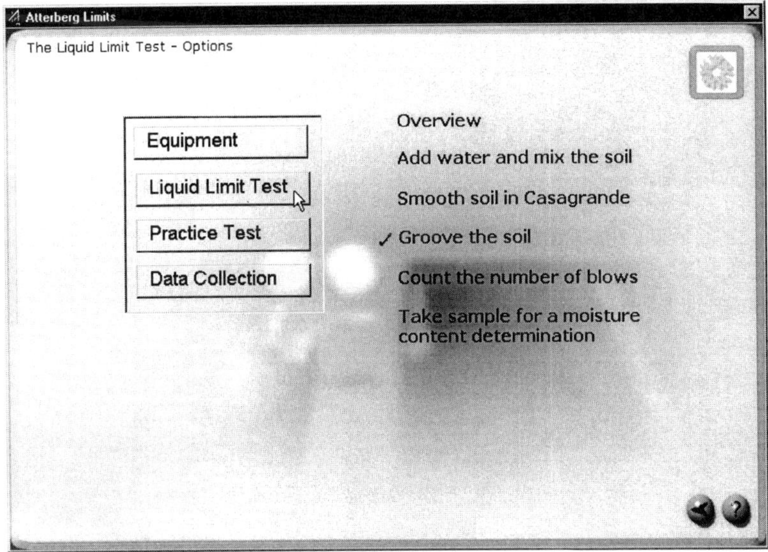

Figure 7. The Liquid Limits Test screen

The options, "Liquid Limit Test" from the screen shown in Figure 5 will take the user to the screen shown in Figure 7, which illustrates sample preparation and the test process using an animated sequence. From here, the student can follow the real sequence of the test, or alternatively, jump around.

This portion allows the students to learn the test sequence, which can then be examined by actually trying the "Practice Test". Figure 8 shows a frame from an animated sequence where the liquid limit device is being turned and the student attempts to keep track of the number of blows, or rotations. At the completion of the test, the student is presented with a series of cups, Figure 9, and is asked to select the groove which has closed 13 mm. The student can verify the 13mm closure by dragging the scale over the cup to make sure that there is indeed closure over the required 13 mm.

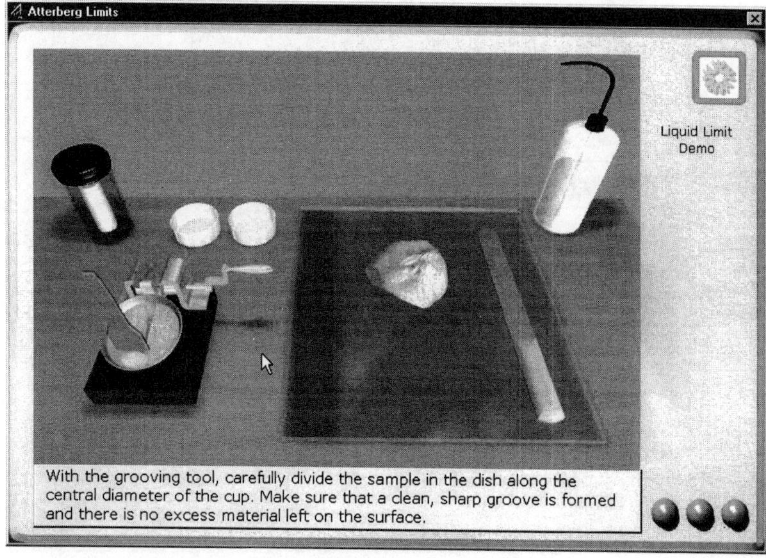

Figure 8. A single frame from the Liquid Limit animation

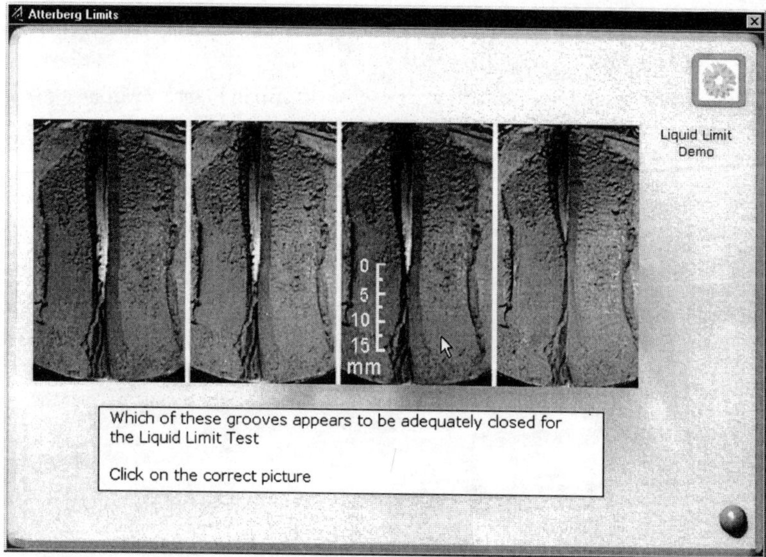

Figure 9. Interactive question at end of liquid limit "jarring" of cup

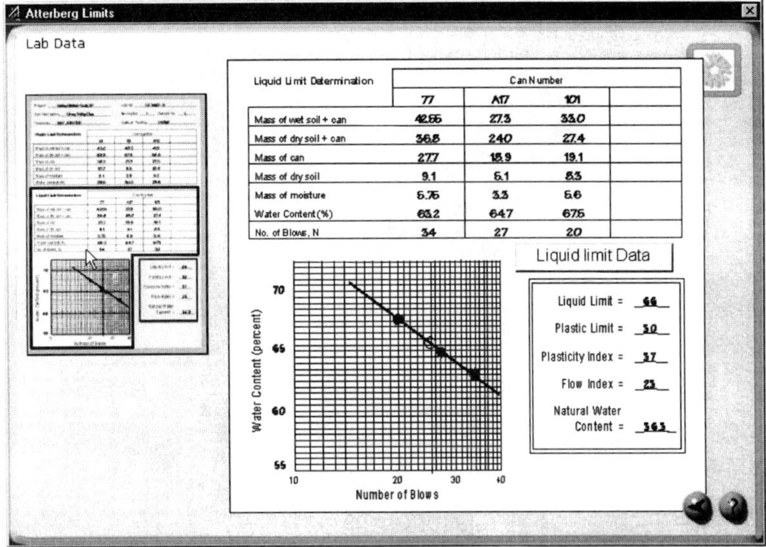

Figure 10. Example of raw and final data for the Liquid Limit test

One of the most important parts of the laboratory test is the data collection and reduction phases. To illustrate this, the "Data Collection" option takes the user to the screens shown in Figure 10. This figure shows an example of "raw" and computed information collected during the test and recorded on the supplied data sheets for the Liquid Limit test. By clicking on portions of the thumbnail sketch on the left, the data sheet is expanded and several explanations are offered concerning each part.

The other part of this module relates to the Plastic Limit test. A view of the typical equipment used for this test is shown in Figure 11 and is accessed from a screen similar to Figure 5, which is for the liquid limit test. To illustrate the actual testing part for the Plastic Limit test, video clips are used to show various stages that range from collecting the fine fraction (using sieves) to rolling the threads until they reach the plastic limit. A frame showing the *second* author (expertly) rolling the threads is shown in Figure 12.

SUMMARY

In this short paper, we have introduced the multimedia approach being used to enhance the learning experience of civil engineering students enrolled in their first course in Geotechnical Engineering or for programs with limited laboratory facilities. The modules may also be used by non-civil engineering programs and for training technicians who are not familiar with the common geotechnical engineering tests.

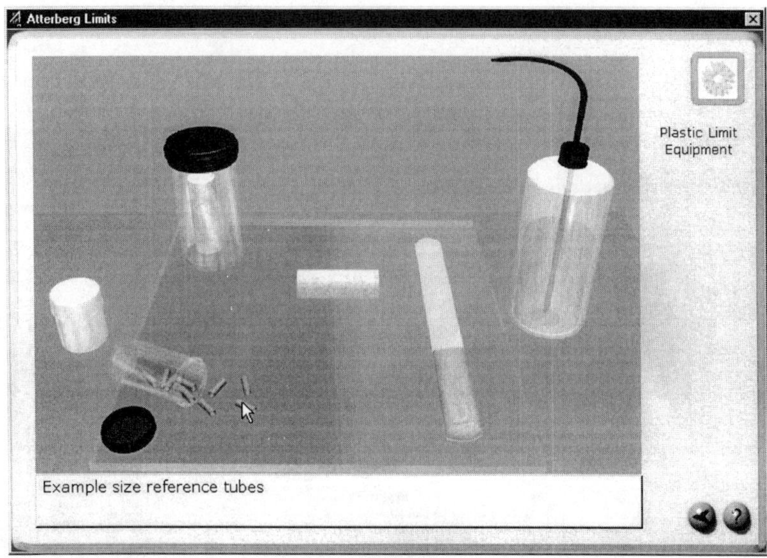

Figure 11. Equipment for the Plastic Limit test

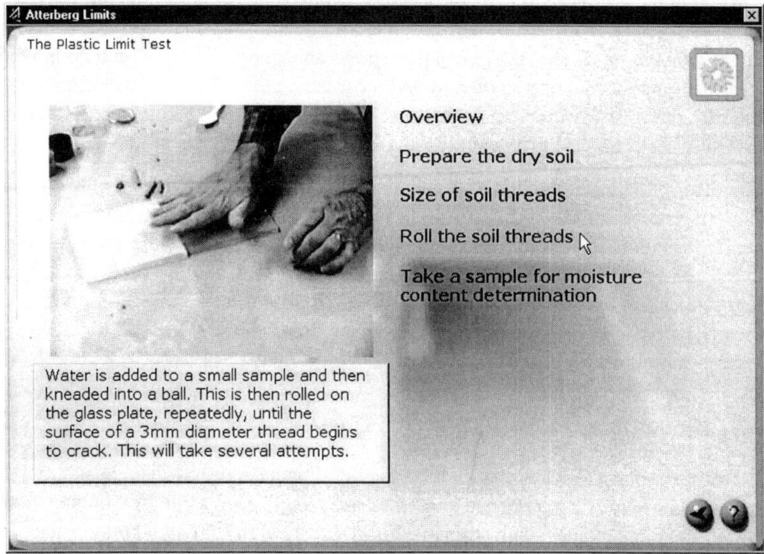

Figure 12. Part of video clip showing the "rolling" action

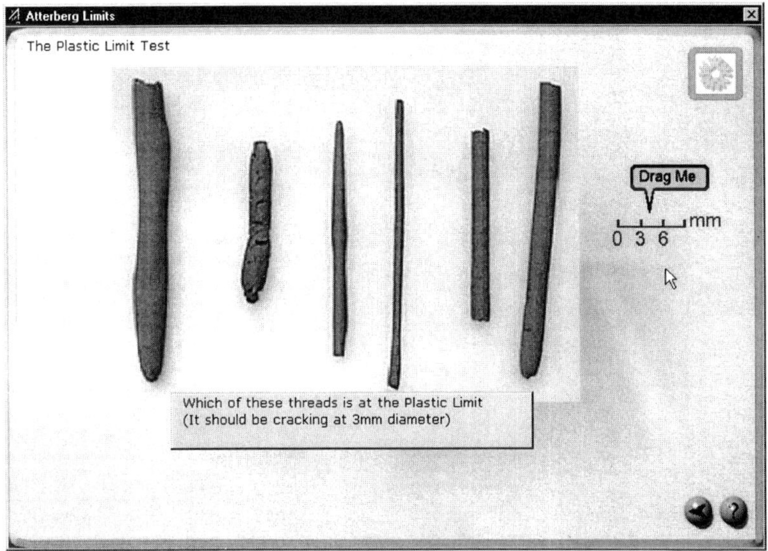

Figure 13. Display of a variety of threads close to plastic limit

The computer assisted approach should improve the capability of the students in performing the actual test, and hopefully, increase their overall awareness to the difficulties associated with the testing of soils.

ACKNOWLEDGEMENTS

We would like to acknowledge the wonderful work of Rob Barnes who created many of the graphics, animations, video-clips and figures. Of course, none of this would have been possible without the foresight of the Idaho State Board of Education that provided financial support under their "Technology Incentive Grant" program and the resources provided by the Civil Engineering Department at the University of Idaho.

Footnote

After the paper had been prepared and submitted, Professor R. D. Holtz, University of Washington, Seattle, WA, pointed out that the threads shown in Figure 13 are too long and do not meet common specifications. We concur with this observation, and promise to fix this example view in the Atterberg Limits module.

A VIRTUAL TRIAXIAL TEST COURSEWARE

Muniram Budhu[1]

Abstract

The purpose of this paper is to present a web-based multimedia courseware to simulate a triaxial test. The benefits of this courseware include enhancing learning, retention and transfer of learning, improving the connection between theory, experiments and applications, making laboratory simulations possible at remote locations, and use of effective pedagogical strategies. Interactions are coded in the courseware to recall prior knowledge, to stimulate learning and to test learning outcomes. The courseware simulates all the procedures that a student normally performs in a laboratory class and more. Each student has his or her own virtual apparatus to conduct not only the standard test but also a variety of tests that cannot normally be performed in a regular laboratory course. Preliminary evaluation shows that the virtual triaxial test enhances learning about the test, testing procedures and interpretation of the results.

Introduction

In geotechnical engineering, undergraduate (mostly at junior level) and graduate students conduct an assortment of soil tests to fulfill the course requirements. Many of these soil tests are time consuming and geotechnical laboratories require significant financial and technical resources to update and maintain. These resources have become scarce resulting in a range of problems in soil laboratory instructions. The investigator contacted colleagues at several institutions and some of the problems of soil laboratory instructions cited are as follows.

- Lack of testing of pre-requisite knowledge
- Students do not make the connection between theory, experiment and application.

---

1 Professor, Department of Civil Engineering and Engineering Mechanics, University of Arizona, Tucson, AZ 85721, USA

- Some students view experiments as a chore
- Some students do not participate in the experiments and consequently do not get hands on experience of setting up the apparatus and conducting the tests because of inadequate number of apparatuses and lack of laboratory space
- Students do not perform the whole test because of time constraints
- Equipment are out-dated and funds for replacement are inadequate or non-existent
- Some institutions at remote locations do not have laboratories
- Lack of teaching aids to vividly explain the intricacies of the various laboratory tests and to excite students to seriously conduct the experiments.
- Practically impossible to investigate 'what-if' situations.

With the advent of computer-based technology (CBT), a large number of courses have been and are being developed for web-based instructions. It has become relatively easy to develop web-based course materials for courses that do not have laboratory components. However, technology is currently available that allows for the development of virtual laboratories. It is now possible to develop laboratory courseware that can simulate very sophisticated and costly equipment and reach a larger section of the student population. In addition, the courseware can be very valuable in enhancing learning, evaluating learning outcomes and solve most of the problems outlined above. The intention behind this paper is to present the development of a virtual triaxial test that can potentially benefit all geotechnical engineering undergraduates and can enhance learning and retention.

Conventional Triaxial Test

The triaxial test is used to study the stress-strain behavior of soils and to determine shear strength parameters to design foundations. The basic apparatus consists of a loading frame, a pressure chamber (cell) enclosing the sample and other appurtenances. The triaxial apparatus has become very sophisticated with computer controlled loading systems and data acquisition systems (GDS Instruments Ltd, 1998). The cost of these newer apparatuses is often out of the budgetary reach of many educational institutions. A schematic of a conventional triaxial apparatus is shown in Figure 1.

Many different types of tests can be conducted in this apparatus but, usually, only one type and one test sample per student or group is used in laboratory instruction because of time constraints.

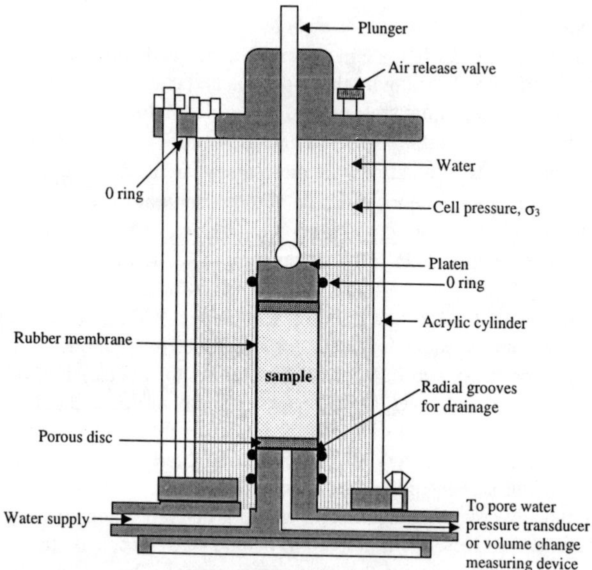

Figure 1 A schematic of a typical triaxial cell.

The usual instructional procedure consists of

1. Preparing one (usually) or more samples for testing.
2. Place the soil sample in the pressure chamber and assembly the apparatus.
3. Apply back pressure to saturate the soil, if necessary. In many cases, students do not apply back pressure to fine grained soils because of the long time required to achieve the desired level of saturation.
4. Apply confining pressures and allow the soil to consolidate.
5. Apply vertical stress to shear the soil.
6. Record vertical displacement, pore water pressures or volume change indicated by dials and manometers.
7. Plot the results and determine soil parameters, e.g., angle of friction, etc.

The triaxial apparatus is very versatile and students rarely get to explore the full range of test options. In many institutions, only one rather than the desired three tests are normally performed by the students because of time, equipment and space constraints. In one type of test, the consolidated drained test on fine-grained soils, the experiment on one test sample can last for two or more weeks and students do not conduct such test although it is taught in a regular geotechnical course.

Virtual Triaxial Test

The virtual triaxial test is a multimedia web-based courseware that is intended to replicate all the procedures that a student will perform in a real laboratory setting. In addition, modern and effective pedagogy to enhance learning and retention is included. The instructional interactions coded into the courseware stimulate recall of prior knowledge, let each student conduct virtual triaxial tests in his or her apparatus and determine the relevant soil parameters, test key principles on the subject matter and provide immediate feedback. Some key features of the virtual triaxial test courseware are as follows.

1. A simulated apparatus with functional gauges and other instrumentation in a virtual laboratory. Usually, instructional apparatus have dial gauges to monitor displacements. However, many commercial laboratories and modern devices have electronic displacement devices. These modern devices are simulated in the courseware to expose students to the latest technology.
2. Each student has his or her own virtual apparatus and his or her own soil sample.
3. A student can assemble the apparatus, prepare a test sample, place the sample in the apparatus, add loads, witness the deformation of the sample, interpret the results, and determine soil parameters from the test results.
4. The courseware tests a student's prior knowledge that is relevant to the triaxial test and quiz students during the virtual laboratory exercise to evaluate learning outcomes.
5. Students are provided with immediate feedback, performance evaluation and help.
6. Several 'what-if' situations are coded into the courseware. For example, if a drained test was conducted, the student has the option to conduct an undrained test on the same soil and or retest the soil at a different over-consolidation ratio.

The modified cam-clay model is used to predict the stress-strain response. The water content and over-consolidation ratio of a simulated soil are randomly generated within practical ranges of values. The parameters for the modified cam-clay model (Roscoe and Burland, 1968) are then obtained from empirical relationships proposed by Azzouz et.al. (1976), Iizuka and Ohta (1987), Wood and Wroth (1978), and Wood (1990). Perturbation functions are randomly introduced to alter the modified cam-clay predictions so as to mimic real soil responses. For example, some stress points in a stress-strain plot are shifted from the predicted values to make the stress-strain curve not as smooth as predicted by the modified cam-clay model. Several commercial software programs including Authorware, Director, Visual Basic, Visual C, C++, Premiere and Photoshop were used to develop various elements of the simulations.

The instructional methods and steps follow guidelines for effective learning as proposed by Gagné (1985) and Clark (1989). These include: gaining attention, informing students of the objective, stimulating recall of prior knowledge, presenting the stimulus, providing learning guidance, eliciting performance, providing feedback, assessing performance, enhancing retention and transfer of learning.

Virtual triaxial soil tests have been proposed by others (Arduino et. al., 1997, Davies et. al., 1997). In the virtual triaxial test presented by Arduino et. al. (1997), the apparatus is made of geometrical elements that does not mimic a real triaxial apparatus. A blackboard schematic is used to display test results predicted by using the hyperbolic model of Duncan and Chang (1970). LabSim (Davies et. al., 1997) used two-dimensional line elements to represent the triaxial apparatus and the modified cam-clay model to simulate a soil sample under loads. In these virtual triaxial tests, students do not prepare the sample, do not set up the apparatus and pre-requisite knowledge is not tested. It is unclear if opportunities exist for students to investigate "what-if" situations.

Examples of Some Key Elements of the Virtual Triaxial Test

The implementation of some of the key instructional elements is described below.

Gaining Attention: A description of the test, why the test is necessary and the presentation of an image of shear failure is used to gain students' attention.

Objective: The objective of the test is presented with information on navigating the courseware and how the scoring will be done.

Stimulating Recall of Prior Knowledge: The objective of the interactions under this instructional element is to test the students' mastery of the prerequisite skills before proceeding with the experiments. The student is given a soil profile and is requested to calculate the vertical effective stress and the void ratio for his or her soil sample. Realistic soil properties are randomly generated so that each student has a different soil profile. Figure 2 illustrates how this is done for the vertical effective stress.

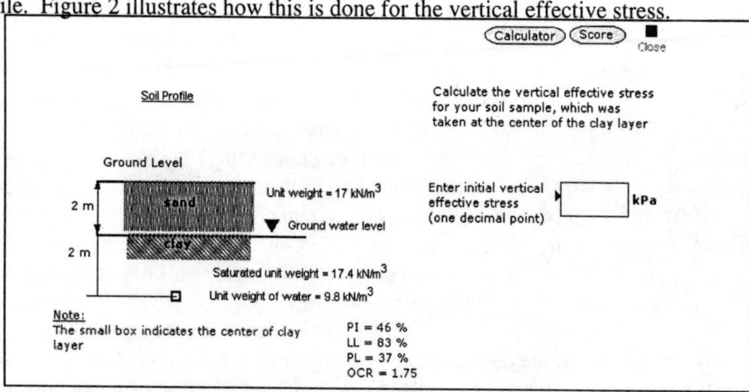

Figure 2 Questions to test prior knowledge

Present the Stimulus and Provide the Learning Guidance

Assembling the Triaxial Cell.

The student is presented with a complete apparatus and identifies important parts of it (apparatus). Then he/she sets up a sample in the triaxial cell and assembles the apparatus using drag and drop interactions. Figure 3 illustrates a screen shot of one stage of the setting up procedure.

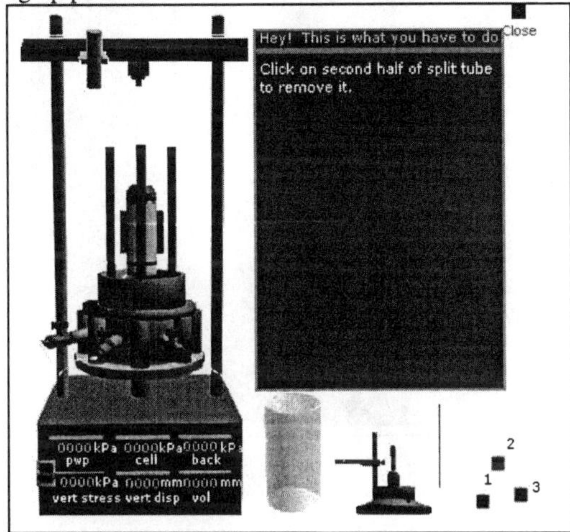

Figure 3 Setting up a sample for testing and assembling the apparatus.

Cell and Back Pressures

The student opens a valve to fill the cell with water and then applies a desired cell pressure. Options are then presented for the use of back pressure with information on the current degree of saturation. If the student opts to apply back pressures, he/she specifies a value of the back pressure and opens a valve to apply the desired value. Back pressure elapsed times are displayed and the student checks whether saturation is achieved by applying a small cell pressure and noting the change in excess pore water pressure. Then he/she calculates the degree of saturation.

Isotopic and K_o Consolidation

Options are provided for conducting an isotropic or a K_o-consolidation test.

The student specifies and applies the desired cell and axial pressures. Guidance is provided on the values to be applied. A screen shot of a sample undergoing consolidation is shown in Figure 4.

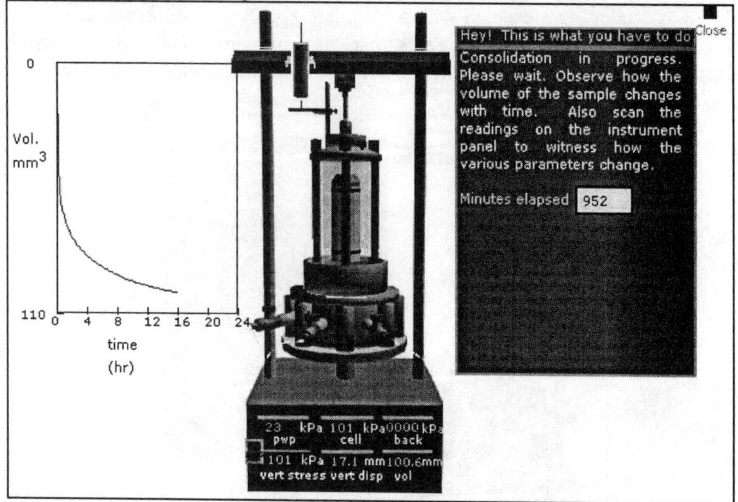

Figure 4 Screen shot of consolidation in progress.

Triaxial Compression

After consolidation is completed, the student carries out a standard triaxial compression test. The over-consolidation ratio of the soil is randomly generated with values within a practical range. Thus each student has a soil with a different over-consolidation ratio. The student then selects whether he/she wants to conduct a drained or undrained test and then switches the motor on to apply axial stresses. Graphs of deviatoric stress versus axial strain and volumetric strain versus axial strain (drained tests) or excess pore water pressure versus axial strain (undrained tests) are plotted during the application of the axial stress. At the end of the first test, the student is prompted to specify different cell and axial pressures (confining stress) for a second test and the stress-strain graphs are shown for the new confining stress. The process is repeated for a third test. Stress strain graphs plotted up to the second test are shown in Figure 5.

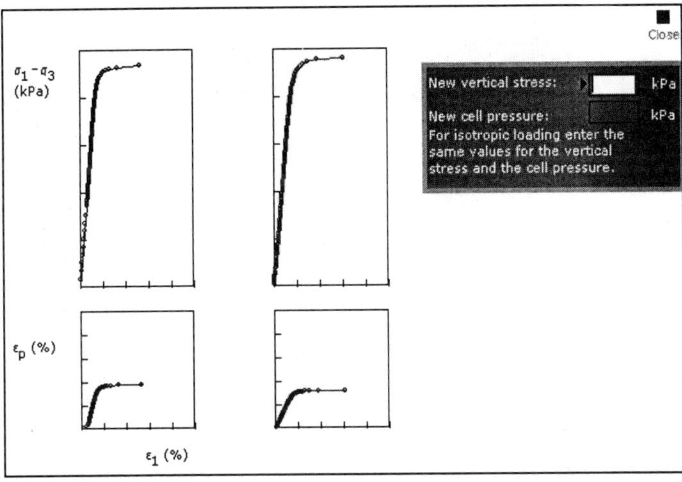

Figure 5 Stress-strain graphs for tests 1 and 2.

Interpretation of the Results

From the three tests the student extracts the peak and critical deviatoric stress and the corresponding volumetric strain or excess pore water pressures. Each student then calculates the secant modulus for one of the tests, the minor and major principal total and effective stresses and the friction angles for all tests. A table listing the extracted stress values for a drained test is shown in Figure 6. These extracted values are used to plot three Mohr circles and then the student determines the peak and critical state friction angles, the undrained shear strength, the failure stresses and the stresses on any desired plane. An illustration of the interpretation of the results using Mohr circle is shown in Figure 7.

These are your results.
Calculate and enter the major principal effective stress

Test	σ'_3 kPa	$\sigma_1 - \sigma_3$ kPa	σ'_1 kPa
1	101	83.7	
2	200	353.4	
3	300	533.5	

Figure 6 Table of extracted stress values from which the student is prompted to

calculate the major principal effective stress for a drained test.

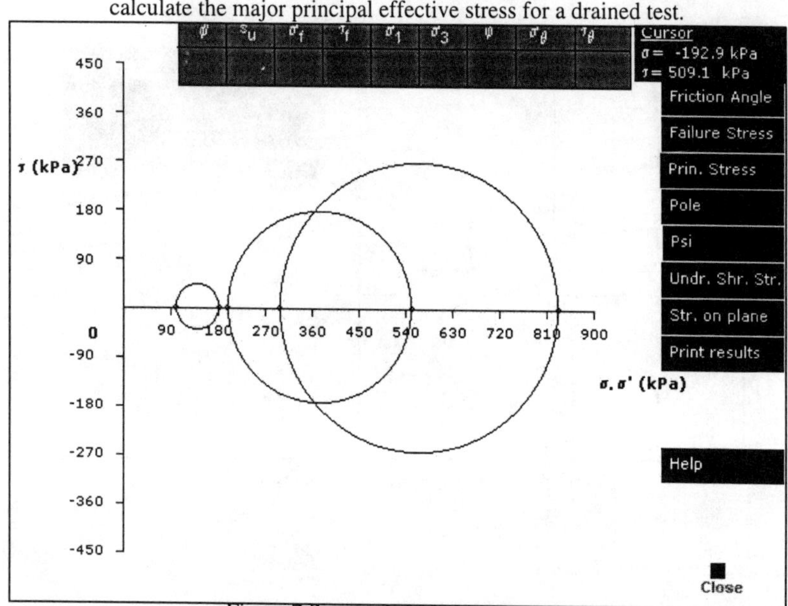

Figure 7 Screen shot of Mohr Circles

What-if Situations

What-if situations are coded into the virtual triaxial test courseware to allow students to learn changes in soil responses and interpretation of the results if the drainage conditions were changed or if the consolidation of the soil were different. For example, if the student conducted a drained test, he/she can opt to conduct an undrained test on the same soil or specify a different over-consolidation ratio and can witness the changes in soil response without setting up the apparatus.

Elicit Performance and Provide Feedback

The internal process of responding is a critical part of the learning process. Each step a student takes in an interaction is monitored and if he or she gets a correct answer to a question, a congratulatory message appears to that effect. If he/ she is wrong, he/ she is prompted to try again. Each attempt to answer the question correctly is recorded and a tally is taken at the end of the interactions. After the third attempt, the solution to the question is presented.

Evaluation

Preliminary evaluation of the virtual triaxial courseware was conducted at the University of Arizona. Students were asked to fill in a questionnaire consisting of ten statements. One of the statements is "I am able to conduct experiments and/or otherwise experience situations that I would not be able to otherwise." The students could respond with one of the following – strongly agree, agree, disagree, strongly disagree and no basis for judgement. The answers from thirty-six students were 28% strongly agree, 56% agree, 11% disagree and 5% strongly disagree.

Conclusions

The virtual triaxial test courseware can be used as a stand-alone laboratory module or can be used to supplement the actual laboratory tests to enhance learning, to extend the range of testing, to integrate basic theoretical concepts/principles and lab tests, to test and transfer fundamental principles. It can also be used as a distance learning module. The preliminary evaluation shows that it benefits 84% of the students in a particular class at a particular institution. More widespread evaluation is needed to determine the effectiveness of the virtual test in geotechnical engineering courses.

References

Azzouz, A.S., Krizek, R. J. and Corotis, R. B. (1976) Regression analysis of soil compressibility, Soils and Foundations, JSSMFE, Vol 16, No. 2, pp.19-29.

Clark, R. C. (1989) Developing Technical Training, Buzzards Bay Press, Phoenix, AZ.

TG Davies, K McColl and S McSporran (1997) LabSim, http://www.civil.gla.ac.uk

GDS Instruments Limited (1998) New PC-controlled standard stress path triaxial testing system, GDS catalog, Surrey, UK

Gagné, Robert (1995) The conditions of learning, Holt, Rinehart and Wilson, NY.

Iizuka, A. and Ohta, H. (1987) A determination procedure of input parameters in elasto-viscoplastic finite element analysis, Soils and Foundations, JSSMFE, Vol. 27, No. 3, pp. 71-87.

Roscoe and Burland (1968) On the generalized stress-strain behaviour of 'wet' clay', in J. Heyman and F. A. Leckie (eds), Engineering plasticity, Cambride University Press, pp. 535-609

Wood, D. M. and Wroth, C. P. (1978) The correlation of index properties with some basic engineering properties of soils, Canadian geotechnical Journal, Vol. 15, No. 2 pp. 137-145.

Wood, D. M. (1990) Soil behavior and critical state soil mechanics, Cambridge University Press, Cambridge

Acknowledgement

The development of the virtual triaxial test was partly funded under grant DUE-9950906 from the National Science Foundation (NSF). The author is grateful to NSF program director, Dr. Eric Sheppard, for his encouragement and support.

HANDS-ON UNDERGRADUATE GEOTECHNICAL ENGINEERING

By

Gregg L. Fiegel,[1] Associate Member, ASCE, and Jay S. DeNatale,[2] Member, ASCE

ABSTRACT: A primary goal of the geotechnical engineering program at Cal Poly, San Luis Obispo is to integrate field and laboratory experiences into the undergraduate curriculum using a hands-on approach. The faculty feel strongly that students must have a complete understanding of geotechnical field and laboratory testing procedures if they are to appreciate the judgment required in geotechnical analysis and design. A unique aspect of the Cal Poly geotechnical engineering program is that undergraduate students have the opportunity to gain extensive experience with field testing equipment. Several undergraduate courses incorporate field testing, thus allowing students to work with the program's trailer-mounted cone-penetrometer testing system and truck-mounted drilling rig. Described within this paper is Cal Poly's "hands-on" geotechnical engineering program. Highlighted are experiences that the faculty have had incorporating field and laboratory testing procedures into the undergraduate curriculum. In addition, a brief discussion focuses on technical and financial support provided by industry. This support has greatly enhanced the capabilities of the program.

INTRODUCTION

Defining the characteristics of a subsurface profile is a challenge that is unique to geotechnical engineering. It is important, therefore, that aspiring geotechnical engineers have knowledge of site exploration strategies, sampling techniques, and laboratory testing procedures before entering professional practice.

[1] Associate Professor, Department of Civil and Environmental Engineering, California Polytechnic State University, San Luis Obispo, CA 93407.

[2] Professor, Department of Civil and Environmental Engineering, California Polytechnic State University, San Luis Obispo, CA 93407.

In most undergraduate programs, students are required to complete one introductory laboratory course where they learn the basics of geotechnical testing. However, when field testing and soil sampling topics are introduced, many undergraduate geotechnical programs rely on textbooks, videotapes, and/or very brief demonstrations. This approach seldom allows the student to acquire a true feel for the thought and work involved in a typical site investigation.

A primary goal of the geotechnical engineering program at California Polytechnic State University, San Luis Obispo (Cal Poly) is to integrate field and laboratory experiences into the undergraduate curriculum using a hands-on approach. Seven of the eleven geotechnical engineering courses that are offered incorporate some form of field and/or laboratory experimentation. The faculty feel strongly that students must have a complete understanding of geotechnical field and laboratory testing procedures if they are to appreciate the judgment required in geotechnical analysis and design.

A unique aspect of the geotechnical engineering program at Cal Poly is the fact that undergraduate students have the opportunity to gain extensive experience with field testing equipment. Several undergraduate courses incorporate field testing, thus allowing students to work with the program's trailer-mounted cone-penetrometer testing system and truck-mounted drilling rig. Typically in these courses, start-to-finish designs are tackled where the students plan a geotechnical field investigation, complete the investigation, interpret the data, and perform the necessary analyses.

Described within this paper is the hands-on geotechnical engineering program currently in place at Cal Poly. Detailed are experiences that the program faculty have had incorporating field and laboratory testing procedures into the undergraduate curriculum. In addition, separate sections focus on industrial participation in the curriculum and on "service-learning" opportunities that have been developed. Finally, a brief discussion focuses on financial support solicited from industry that has helped to greatly enhance the capabilities of the program.

BACKGROUND

California Polytechnic State University, San Luis Obispo

Cal Poly, San Luis Obispo is one of the twenty-three (23) campuses of the California State University, the nation's largest four-year university system. Cal Poly's primary mission is undergraduate education, and it specializes in preparing its 15,000 students for careers in applied professional and technical fields. Cal Poly's College of Engineering has been consistently identified as one of the nation's top state-funded engineering programs. Its "learn by doing" motto is taken very seriously by the students, faculty, and staff. Undergraduates are required to take numerous laboratory classes as well as a two-quarter, capstone senior research/design project. The hands-on experience gained in these activities stimulates creativity and self-discovery while preparing students for the rigors of industry.

Civil and Environmental Engineering Department

The Civil and Environmental Engineering (CEEN) Department is housed within the College of Engineering. The Department provides approximately 760 undergraduates with a theoretically sound, practice-oriented education. Taught by nineteen tenured and tenure-track faculty and several local professionals, the program consists of 208 quarter units with courses in each of the five traditional civil engineering sub-disciplines: environmental, geotechnical, structural, transportation, and water resources engineering. Laboratories account for 21 units of the required curriculum.

Geotechnical Engineering Program

The geotechnical engineering program administered by the CEEN Department is directed by two full-time faculty and one part-time lecturer. All are registered Professional Engineers in the State of California. The geotechnical program has been recognized as one of the top undergraduate programs in California. Cal Poly geotechnical engineering students have received approximately 10% of all International Association of Foundation Drilling (ADSC) scholarships awarded since 1990. In addition, Cal Poly students have received the California Geotechnical Engineers Association (CGEA) annual Outstanding Graduating Senior Award five of the past ten years (1990, 1991, 1992, 1995, and 1999).

Cal Poly graduates with a geotechnical background are well prepared for engineering practice or graduate school. In recent years, graduating seniors have gone to work for many of the major West Coast geotechnical engineering firms including Woodward-Clyde, Law/Crandall, Dames & Moore, and Treadwell & Rollo. Graduate schools attended by Cal Poly students include the University of California (Berkeley, Davis, and Los Angeles), the University of Washington, the University of Colorado, and Virginia Polytechnic Institute, among others. On average, about four graduating seniors decide to enter a geotechnical graduate program each year.

GEOTECHNICAL CURRICULUM

The content of a particular course depends, in part, on course objectives. Within the geotechnical engineering program, course objectives have been set to ensure: (1) the development of a solid base for advanced undergraduate and/or graduate coursework and professional practice in geotechnical analysis and design; (2) maximum competence in the specific types of geotechnical analysis and foundation design problems that are identified in the Examination Specifications for the National Council of Examiners for Engineering and Surveying (NCEES) Principles and Practice of Civil Engineering (PE) Examination (NCEES 1995); (3) fluency with the process of engineering design, as defined by the Accreditation Board for Engineering and Technology (ABET 1994); and (4) fluency in graphical, written, and verbal communication, as required for professional practice.

A total of eleven geotechnical engineering courses are available for civil and environmental undergraduate students interested in acquiring knowledge and basic skills related to the above four areas (see Table 1). These courses are well supported by members of the professional community who serve as guest lecturers, organize field trips, and participate in the development and evaluation of student design projects.

Undergraduate civil engineering students at Cal Poly are required to take one course and one laboratory in basic geotechnical analysis (CE 381 and CE 382). Environmental engineering students must take CE 381. A civil or environmental engineering student interested in geotechnical engineering may then choose from nine geotechnical electives offered by the department. Shallow and deep foundation design, slope stability analysis, geotechnical earthquake engineering, environmental geotechnology, and lateral earth pressure theory are among the topics covered in the elective courses.

Table 1. Geotechnical Engineering Courses Offered at Cal Poly

Course Number	Title	Quarter Units	Laboratory Component
CE 381	Introduction to Geotechnical Engineering	4	N
CE 382	Geotechnical Engineering Laboratory	1	Y
CE 481	Analysis and Design of Shallow Foundations	4	Y
CE 482	Conventional Subsurface Exploration	4	Y
CE 483	Environmental Geotechnology	4	N
CE 581	Advanced Geotechnical Engineering	4	Y
CE 582	Advanced Geotechnical Testing	4	Y
CE 583	Geotechnical Earthquake Engineering	4	Y
CE 584	Lateral Support Systems	4	N
CE 585	Slope Stability Analysis	4	Y
CE 586	Analysis and Design of Deep Foundations	4	N

High undergraduate student enrollments are typical in the geotechnical courses. Figure 1 illustrates the enrollment trend in the geotechnical area over a twelve year period. The figure shows that an average of approximately 300 seats are filled in geotechnical analysis and design courses each year.

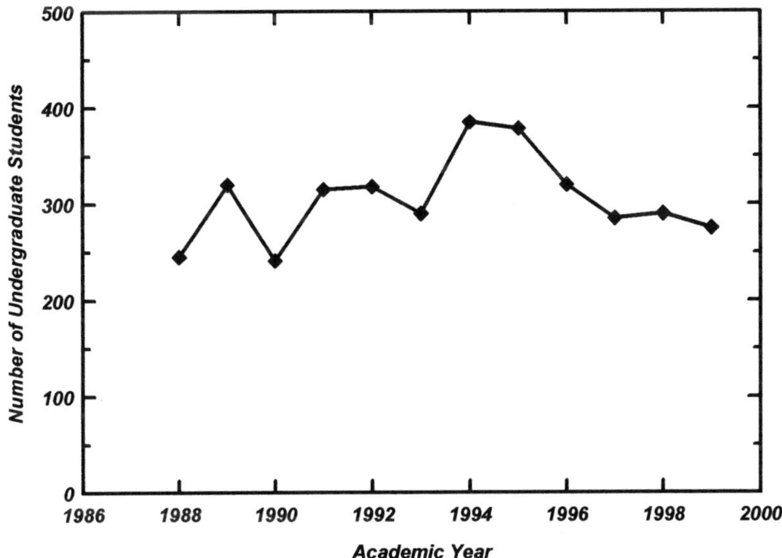

Figure 1. Enrollment in Geotechnical Engineering Courses at Cal Poly

EXAMPLES OF THE HANDS-ON APPROACH

It is important to remember that course content also depends, to a degree, on the abilities and orientations of the target audience (i.e. course delivery is extremely important). At Cal Poly, the geotechnical engineering faculty believe that course delivery should contain a substantial amount of hands-on work. This hands-on work is accomplished through laboratory instruction, field instruction, industry sponsored projects, and service learning projects.

Laboratory Instruction

The engineering curriculum at Cal Poly is laboratory intensive. As summarized in Table 1, seven of the eleven geotechnical engineering courses have a laboratory component. These geotechnical laboratories are commonly taught in the field, in one of two well-equipped teaching laboratories, or at an available computer room. During a typical ten-week quarter, a laboratory course will have six to eight organized lab sessions each lasting anywhere from two to three hours. The lab sessions, designed with input from industry, simulate situations that students will eventually see in professional practice. Required laboratory reports or term projects are modeled after project reports provided by outside geotechnical engineering consultants.

Field Instruction

Since 1986 the geotechnical engineering program has owned a trailer-mounted cone penetrometer testing system. The system was originally supplied with two cones, a flat-plate dilatometer, and an electronic data acquisition system. Recently, the data acquisition system was upgraded using funds provided by a National Science Foundation Division of Undergraduate Education grant (DUE #9751318). The new Vertek-manufactured system includes two piezocones and a seismic cone.

Traditionally, the cone penetrometer and flat-plate dilatometer testing systems have been used in the Advanced Geotechnical Testing (CE 582) course. Students in this course spend much of their time in the field performing cone penetrometer and dilatometer tests under the direct supervision of faculty. Four- or five-person laboratory groups are used so that each student receives hands-on experience with the equipment. Data and soil samples retrieved during field exploration are examined as part of detailed design projects. The design projects teach the students how to plan and complete subsurface investigations, interpret field test data, and design simple foundation systems.

Student evaluations of the geotechnical field testing course indicate that it is very well received. Some students even name the course as their favorite at Cal Poly. The last time this course was taught the instructor received an overall rating of 3.9/4.0 during student evaluations. Some example student comments taken directly from these same evaluations are listed below.

"I learned a lot in this class."

"Nice to learn about field methods that we are not exposed to in the classroom."

"Great course, should be offered in the future."

"I finally understand CE 381 and CE 481, thanks."

Based on the above comments and individual conversations with the students, the hands-on, start to finish approach appears to be the most attractive element of the geotechnical testing course (CE 582).

In 1997 Foremost/Mobile of Indianapolis, Indiana agreed to provide Cal Poly with a completely refurbished Mobile B-53 Explorer drill rig at a substantially reduced cost (see Figure 2). Funds required to purchase the rig and tooling were obtained from the National Science Foundation's Division of Undergraduate Education, numerous Cal Poly geotechnical engineering alumni, the West Coast Chapter of ADSC, the CEEN Department, the College of Engineering, and Cal Poly.

The rig was mounted on a truck donated by Anderson Drilling of San Diego, California, a prominent member of the ADSC West Coast Chapter. Tooling for the rig was provided by Foremost/Mobile through Drilling and Trenching Supply, also a member of ADSC (Fiegel 1998).

Figure 2. Cal Poly's Truck-Mounted Drill Rig

The newly acquired drill rig has significantly enhanced the education of Cal Poly's civil and environmental engineering undergraduates by providing them with a means of acquiring hands-on training in conventional subsurface exploration and field testing. The rig can be used to perform either hollow-stem auger or mud rotary drilling. Split-spoon and thin-walled sampling tools are available. Drive samples are obtained using either a safety hammer or an automatic trip hammer. Overall, the drill rig, in combination with the cone penetrometer system, provide Cal Poly students with unparalleled and unprecedented exposure to all aspects of real-world geotechnical investigation, analysis, design, and construction.

To date, the drill rig has been used in conjunction with several courses and projects. Some examples of how the rig is used in a typical academic year are listed below.

- Site exploration and sampling methods are briefly demonstrated during sessions of CE 381. The primary objective of these sessions is to stimulate interest in geotechnical engineering among the juniors of the civil and environmental engineering program.

- The drill rig is used to develop subsurface soil profiles during laboratory sessions of CE 481. Students perform standard penetration tests and obtain undisturbed soil samples using the rig. Results of field and laboratory tests are incorporated into a shallow foundation design project.

- Senior civil engineering students have the option of using the drill rig in their geotechnical-related senior projects. To date, ten senior project students have used the rig. The primary use of the rig in these projects has been as a site investigation tool.

- The drill rig is used as an investigative tool or as a demonstration tool in technical elective courses offered by the program. Courses in which the rig has been used include Conventional Subsurface Exploration (CE 482), Advanced Geotechnical Testing (CE 582), and Geotechnical Earthquake Engineering (CE 583).

The drill rig is used extensively in Conventional Subsurface Exploration (CE 482), a newly developed course that was taught for the first time in the Fall of 1998. In this course, numerous laboratory sessions allow the students compare and contrast different drilling and sampling techniques. However, the primary objective of the course is to teach students how to perform a proper geotechnical site investigation. The results of library research, field testing, and laboratory testing are combined into a term project which is modeled after geotechnical site investigation reports provided by engineering practitioners.

Student evaluations of CE 482 indicate that it is also well received. The first time this course was taught the instructor received an overall rating of 4.0/4.0 during student evaluations. Some example student comments taken directly from these same evaluations are listed below.

"This was the most interesting and exciting hands-on class I have ever taken."

I enjoyed the class and instructor tremendously. The experience of using a drill rig was great."

"This was a very interesting class as well as informative. I feel this is a very important class to take for all geotechnical students."

"Probably the best class I have taken at Cal Poly. Very hands-on, very educational, very relaxed atmosphere, etc."

"Excellent course --- learned how to operate a drill rig and take samples."

Industry Sponsored Class Projects

The geotechnical faculty at Cal Poly serve two sets of clients: the students -- who seek instruction in the theory and practice of civil analysis, design, and construction -- and the professional community -- who seek students with the ability and willingness to "hit the ground running" upon graduation. Continued, direct interaction between these two sets of clients is vital. At Cal Poly, face-to-face communication is fostered by inviting the professional community to participate in the instruction of the students. Approximately ten percent of the lectures in each geotechnical course are delivered by highly respected practitioners. Most of these seminars are case-study type presentations which emphasize the application of analysis and design procedures that have been introduced in earlier traditional lecture-type presentations.

On occasion, visiting practitioners are invited to prepare a realistic design exercise for the students to complete as a homework problem. Normally, these exercises are a limited-scope version of a recent project from the visitor's own practice. In one instance, the students enrolled in the slope stability course (CE 585) were required to respond to a request for geotechnical services that focused on the development of a practical remediation alternative for a local landslide.

The problem statement was delivered to the students on a Saturday morning by three representatives (one Principal and two Project Engineers) from the participating firm (who assumed the role of the client). The problem statement included (1) geologic and topographic maps of the site, (2) a budget, (3) a fee schedule that fixed the hourly rate of each class of personnel (Office Assistant, Lab Technician, Staff Engineer, Geologist, Project Manager, etc.) as well as the unit cost of all conventional field and laboratory procedures, (4) the required scope of work, and (5) the required deliverables. The students were subdivided into five-person "consulting firms" and driven to the site of the landslide (some 20 miles from the campus). After receiving a short 15-20 minute on-site briefing, the students were given two hours to inspect the site and the adjacent residential neighborhood (see Figure 3). The course instructor, the three visiting practitioners, and the site's owner all remained on-hand to respond to student queries.

The competing student-firms were given three days to develop and submit a preliminary report that defined the field investigation (exploratory boring locations and depths), laboratory testing, and field instrumentation programs to be undertaken, should the firm be awarded the contract. Each student-firm was then supplied with the array of geotechnical data (boring logs, lab data, inclinometer profiles) that would be obtained through the field/lab investigation programs specified in the firm's preliminary report. This data was derived from a model of the site that had been developed by an actual consulting firm as part of an actual contract with the site's owner.

The student-firms were then given one additional week to develop and submit a final proposal for remediating the landslide that included (1) a best-estimate of the subsurface stratigraphy, (2) estimates of the safety factor for both the existing and remediated slopes, and (3) a qualitative discussion of the probable environmental impacts of the remediation project. Needless to say, this project was an extremely well-received and worthwhile culminating experience.

Figure 3. A Visit to the Site of a Landslide in the Slope Stability Course

Service Learning Projects

In recent years the faculty of the geotechnical engineering program have begun to use service learning (community service) projects as another form of instruction. To date, these community service activities have been organized through senior design projects and/or the Cal Poly ASCE Student Chapter. However, in the future, the incorporation of service projects directly into the classroom will be considered.

The importance of community service in relation to a student's undergraduate education cannot be understated. Students at Cal Poly working on engineering-related community service projects hone their communication and project management skills. In addition, they are given the opportunity to apply learned concepts and theories to real-life engineering problems. Usually the student volunteers are involved in every aspect of the project from the start to the finish. Hands-on work is the norm. Importantly, students working on service projects realize the personal rewards and satisfaction of giving their own time to help another individual or organization in need. They truly learn what it means to be a civil engineer working in a "people serving" profession.

Several geotechnical-related community service projects have been completed at Cal Poly during the past year. These projects are summarized in Table 2. Included in the table is the project title, the number of students who worked on the project, the number of hours spent on the project, and an indication of whether or not senior project students were involved.

Table 2. Recent Geotechnical-Related Service Projects at Cal Poly

Project Title	Total Number of Students	Total Volunteer Hours	Number of Senior Projects
Geotechnical Engineering for Elementary School Students	14	~ 150	2
Characterization of Soils used in San Luis Obispo Adobe Buildings	12	~ 150	1
Monitoring and Analysis of an Active Landslide at Port San Luis	8	~ 75	2

The projects listed in Table 2 were first organized through the ASCE Student Chapter at Cal Poly, the Society of Civil Engineers (SCE). SCE is a civic-minded organization with a long history of volunteer work in the community. Typically, engineering-related service projects are brought to the club from local civic organizations, community leaders, and/or civil engineers. In the past, SCE has helped in the design and construction of low-income family homes and a small bridge. The elementary school outreach program and the adobe buildings project listed in Table 2 are described in more detail in separate references (e.g. Fiegel et al. 2000). The landslide project listed in Table 2 is described herein to illustrate the level of detail associated with a typical service project.

The landslide project was brought to SCE from a local civil engineer interested in analyzing landslide activity observed along a single lane, coastal access road in Port San Luis, California. This road provides the only access to the Port San Luis Lighthouse, an historical building currently being refurbished by a volunteer group in San Luis Obispo. The group was interested in evaluating several active landslides with the hope of one day stabilizing them against further movement.

Two students directed the landslide project; however, they were supported by six other students within the program. As part of the project, the team first researched the site by examining aerial photographs and reading geologic reports. Data gathered during this phase was used to plan a geotechnical investigation. This investigation included the drilling of two bore holes through two of the active landslides. Cal Poly's drill rig was for this part of the project. The students drilled and logged each bore hole and retrieved several samples. The samples were tested in the laboratory to estimate strength parameters for stability analyses.

After drilling was complete, the students installed slope inclinometers within each bore hole. Inclinometer casing and supplies required for the project were graciously donated by Geokon. The two inclinometers were monitored for a period of several months using equipment loaned to the geotechnical program by Fugro West, Inc. Inclinometer data, lab test results, and field test results were eventually combined so that stability evaluations could be completed for two of the existing landslides. The results of these evaluations were presented to owners of the lighthouse in the form of a report. The hope is that the results provided will be valuable in planning a mitigation technique for each of the active landslides.

INDUSTRIAL SUPPORT FOR THE PROGRAM

The number of scholarships and job offers received by Cal Poly graduates interested in geotechnical engineering indicates that industry supports the hands-on approach. Many practitioners recruit almost exclusively at Cal Poly because of the practice-oriented education that its graduates have received. Interest by the professional community in the Cal Poly geotechnical engineering program has led to tremendous cooperation and support. As discussed, representatives from industry often serve as guest lecturers and participate in the development and evaluation of student design projects. Support is not only technical, however. Recently, representatives from industry have also provided the program with a limited amount of financial resources. The drill rig acquisition is an excellent example.

The drill rig was obtained because of the initiative of the West Coast Chapter of the International Association of Foundation Drilling (ADSC). Two years ago the geotechnical engineering faculty asked the West Coast Chapter for assistance in finding a field exploration rig for the geotechnical engineering program. The West Coast Chapter agreed to help and eventually provided the program with a lead on a refurbished drill rig, a truck carrier, and several thousand dollars in funding (Fiegel 1998). Funding provided by the ADSC West Coast Chapter was combined with contributions from the CEEN Department, the College of Engineering, and Cal Poly and used as matching funds for a National Science Foundation Instrumentation and Laboratory Improvement grant.

Additional required matching funds for the grant were collected during a brief fundraising campaign. Over an eight month period, Cal Poly alumni working the field of geotechnical engineering as well as consulting firms practicing in the geosciences were asked to contribute cash to the drill rig project. Potential donors were first contacted by phone. Then, letters were sent detailing the geotechnical engineering program, the drill rig, and proposed curriculum improvements. The response to this solicitation was overwhelming. During the eight month period, nearly $8,000 in funding was raised which allowed the project to proceed toward completion.

Clearly, there is a history of support for the hands-on geotechnical engineering program being implemented at Cal Poly. This support has continued during the two years that have passed since the completion of the drill rig project. For example:

- The program has accepted annual donations averaging between \$2500 and \$3000 from alumni and industrial sponsors. These funds are used to purchase laboratory supplies and equipment and to sponsor senior projects.

- Last year, Granite Construction of Watsonville, California donated to the program a four-wheel drive pick-up truck valued at \$9000. The truck is used often as a support vehicle during field visits.

- On average, eight to ten organized guest lectures and/or field trips have been organized by engineering practitioners per year.

SUMMARY

A hands-on approach is being used to teach geotechnical engineering to undergraduate students at Cal Poly, San Luis Obispo. The objective of the curriculum is not to train the students to be drill rig operators or laboratory technicians but, rather, to help students acquire the judgment and creativity required in geotechnical engineering. Practice-oriented field and laboratory exercises combined with open-ended design projects are successfully being used to help foster this judgment and creativity. The success has been validated by positive student evaluations and by industry interest in the program.

CONCLUSION

Most academicians and engineering professionals agree that hands-on field- and laboratory-based instruction is an essential component of engineering education (e.g. ASEE 1986; Wankat and Oreovicz 1993). According to Ralph Peck, one of our Nation's most respected and highly-decorated Professional Engineers,

> *"One carefully conducted field observation, together with all of the pertinent data including the soil conditions, is a permanent asset of which the value can never change. It is a permanent part of the experience, not only of the engineer making the tests, but of all engineers who may read the data. It is worth a thousand theories." (Peck 1941)*

> *"Much has been written and much lip service has been paid to the idea of the fundamental variability of natural soil deposits. Yet, the full impact of this variability is almost always lost on an individual whose training has been exclusively along the lines of mechanics, hydraulics, and strength of materials." (Peck 1960)*

Clearly, in geotechnical engineering practice, hands-on experience with soil and soil testing equipment is essential. Given its importance in practice, the authors feel that hands-on experience should be an essential component of any geotechnical engineering educational program, undergraduate or graduate. This is especially important now with the ever increasing popularity and use of computers as tools of the profession. We must remember that, in addition to the computer, essential tools of the geotechnical engineer include the hands, the eyes, and the nose.

It is acknowledged that most university geotechnical engineering programs do not have unlimited access to laboratory and field testing equipment like that found at Cal Poly. The program is unique in that regard. However, this equipment is readily available in practice as engineers and contractors use it on their projects everyday.

Thus, one option for filling the "equipment gap" between industry and academia is partnering. Partnerships between universities and industry help make valuable testing equipment available to students so that they receive the hands-on experience they need before entering geotechnical engineering practice. Such partnerships, which are used extensively at Cal Poly, can be extremely rewarding for faculty and engineering professionals and should be strongly encouraged by the geotechnical engineering community.

ACKNOWLEDGMENTS

Many of the recent improvements to the geotechnical engineering laboratories were made possible by the National Science Foundation's Division of Undergraduate Education through grant DUE #9751318. The authors gratefully acknowledge this support. The authors also acknowledge the financial support provided by the International Association of Foundation Drilling (ADSC) and the numerous Cal Poly alumni who are currently engaged in the practice of geotechnical engineering.

REFERENCES

Accreditation Board for Engineering and Technology, ABET (1994), "*Criteria for Accrediting Programs in Engineering in the United States*," ABET, Baltimore, MD.

American Society for Engineering Education, ASEE (1986), *Quality of Engineering Education Project: Final Report*, ASEE, New York, NY.

Fiegel, G.L. (1998). "Cal Poly Obtains a Truck-Mounted Field Exploration Rig." *Foundation Drilling Magazine*, International Association of Foundation Drilling (ADSC), Dallas, Texas, January, 27-31.

Fiegel, G.L., Elia, V., and Griffith, M. (2000). "Geotechnical Engineering for Elementary School Students." *Proceedings, GeoDenver2000 Specialty Conference*, ASCE, Washington, D.C.

Fiegel, G.L., Vessely, R.S., and Beckmann, J.A. (2000). "Geotechnical Characterization of Adobe Buildings." *Presentation, GeoDenver 2000 Specialty Conference, Session on GeoArcheology*, ASCE.

National Council of Examiners for Engineering and Surveying, NCEES (1995), *Principles and Practice of Engineering: Sample Problems and Solutions in Civil Engineering*, NCEES, Clemson, SC.

Peck, R.B. (1941), "*The Teaching and Practice of Soil Mechanics. A Critical Comparison*," Society for the Promotion of Engineering Education, 6 pp.

Peck, R.B. (1960), "*The Teaching and Practice of Soil Mechanics. A Second Comparison*," Engineering Education, ASEE, Vol. 50, 309-311.

Wankat, P.C. and Oreovicz, F.S. (1993), Teaching Engineering, McGraw-Hill, New York, New York.

Designing a Learning Environment for Geotechnics

Marina Pantazidou[1], Associate Member ASCE

Abstract

Cognitive science theories can now provide the tools to achieve specific instruction goals. The Scaffolded Knowledge Integration (SKI) framework has been identified in this work as particularly promising to offer guidance for the design of effective learning environments for engineering courses. Following the SKI framework, the design involves four main steps. Step 1 articulates the key topics of a course and the associated instructional goals. Step 2 specifies the content of these goals in terms of thinking processes. Step 3 consists of designing learning experiences and instructional tools that will make these thinking processes visible and help students practice them. Finally, Step 4 identifies suitable practices to ensure that the designed learning environment fosters autonomous learning and provides support. This paper discusses the application of the SKI framework to the design of an introductory course in geotechnical engineering and provides examples from work-in-progress on the first three steps of course design.

Introduction

This paper is written for the engineering instructor who is in search of cognitive principles that can be applied to instructional design. Cognitive research results published in the educational literature fall mainly in two categories: overarching guiding principles for instruction (Svinicki 1991; McGilly 1994) or topic-specific learning strategies (Chi et al. 1981; Pirolli and Recker 1994). The former provide valuable information on successful learning practices but little advice on the specifics of translating this information to effective teaching practices. The latter offer detailed advice, but this advice cannot be transferred easily to another subject area. Moreover, the subject areas typically studied are procedurally rich [e.g., computer language instruction (Pirolli and Recker 1994)] or principle driven [e.g., physics (Chi et al. 1981)], with well-defined problems. As a result,

[1] Assistant Professor, Department of Civil and Environmental Engineering, Carnegie Mellon University, Pittsburgh, Pennsylvania 15213

even generalizations from these studies cannot be applied directly to engineering; the engineering instructor needs to identify anew the ways in which instructional theories make sense for a particular engineering subject. This paper has two goals in order to speed up this sense-making task: 1) to familiarize the reader with research on the intersection of cognition and instruction relevant to engineering course design, and 2) to present an example application of a framework for course design, the Scaffolded Knowledge Integration (SKI) framework, by describing ongoing work on redesigning an introductory course in geotechnical engineering.

The Scaffolded Knowledge Integration (SKI) Framework: An Overview

Cognitive science theories can now provide us with the tools to produce concrete instruction results. The Scaffolded Knowledge Integration (SKI) framework, developed by Linn and coworkers at the Graduate School of Education of the University of California at Berkeley (Linn 1995), has been identified in this work as particularly promising to offer guidance for the design of effective learning environments for engineering courses. Unlike most available studies, it provides both a general framework for instructional design, as well as a synthesis of results from specific applications, which include physics, computer language and spatial reasoning courses. Linn describes the Scaffolded Knowledge Integration framework as "a method for encouraging students to develop integrated understanding of a complex domain". To achieve integrated understanding, students need to connect ideas and compare models. Linn defines models broadly "to include heuristics, algorithms, rules-of-thumb, formal mathematical systems, abstract representations and mechanisms". Students come to class with an existing set of such models. The SKI framework offers guidance for designing learning environments and providing support ("scaffolding") to help students modify old models, create new ones and connect the two sets of models.

The four main components involved in the design of a course based on the SKI framework are identified in this paper as Steps 1, 2, 3 and 4. Step 1 articulates the key topics of the course and the associated instructional goals. This requires expertise in the subject matter and teaching experience. Step 2 specifies the content of these goals in terms of thinking processes. This is being accomplished by identifying mental models of students and experts related to these topics, through suitable questionnaires and assignments that will be complemented with interviews. Step 3 consists of designing learning experiences and instructional tools that will make these thinking processes visible and help students practice them. Finally, Step 4 identifies suitable practices to ensure that the designed learning environment, including the tools developed in Step 3, fosters autonomous learning and provides support. This paper situates the specifics of using the SKI framework for instructional design in the context of an introductory course in geotechnical engineering. These specifics will be exemplified with results from work-in-progress on Steps 1, 2 and 3.

It is important to emphasize that the four steps are discussed separately for presentation clarity. This is not to imply that the design process is linear. To underscore this point, Figure 1 provides two counter-examples with schematics of common sequences in building course syllabi (Figure 1a) and designing instructional software (Figure 1b), with assessment focusing on student outcomes. [It is interesting to note that in the case of software design, assessment is typically skewed more towards students' attitudes regarding the learning environment (e.g., frequency of use, ratings of user-friendliness) compared to measures of students' performance.] In contrast, Figure 2 illustrates the iterative SKI course design model and identifies the interrelationships among the four design steps. Figure 2 also indicates that assessment can be integrated with the educational experience, as will be discussed below.

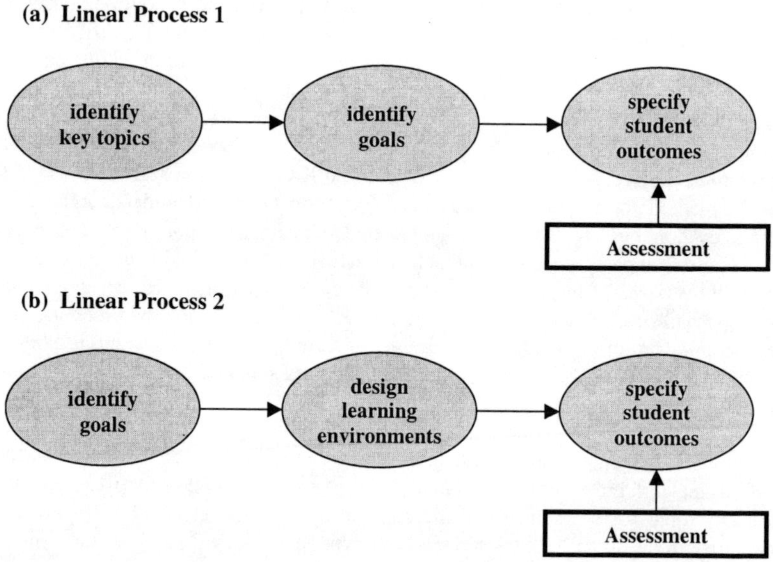

(a) Linear Process 1

identify key topics → identify goals → specify student outcomes ← Assessment

(b) Linear Process 2

identify goals → design learning environments → specify student outcomes ← Assessment

Figure 1. Simplified Schematics of Linear Approaches in: (a) Building a Course Syllabus, and (b) Designing Instructional Software.

Designing an Introductory Geotechnics Course

An introductory course in the engineering behavior of soils aims at familiarizing students with the concepts of soil mechanics needed in geotechnical engineering applications. This first course may be called Soil Mechanics or Introduction to Geotechnical Engineering at different institutions. For brevity, it will be referred to as soil mechanics in the remainder of this paper.

Step 1 - Identify key topics and instructional goals

The first step in course design is identifying the key topics, the associated principles, and the corresponding domain-specific skills; these will determine the instructional goals and the desired student outcomes (see Figure 2). Key topics in applied engineering courses, such as soil mechanics, involve both qualitative descriptions and quantitative analyses of phenomena. As a result, the selected corresponding instructional goals are expressed in terms of declarative ("know that") or procedural ("know how") skills. Research has shown that effective learning requires both types of representations (Reif 1995). The nature of the key topics and skills has direct bearing on the methods used to reveal the thinking contents of these skills (Step 2) and the choices of the instructional tools designed to make this thinking visible (Step 3), as it will be discussed later.

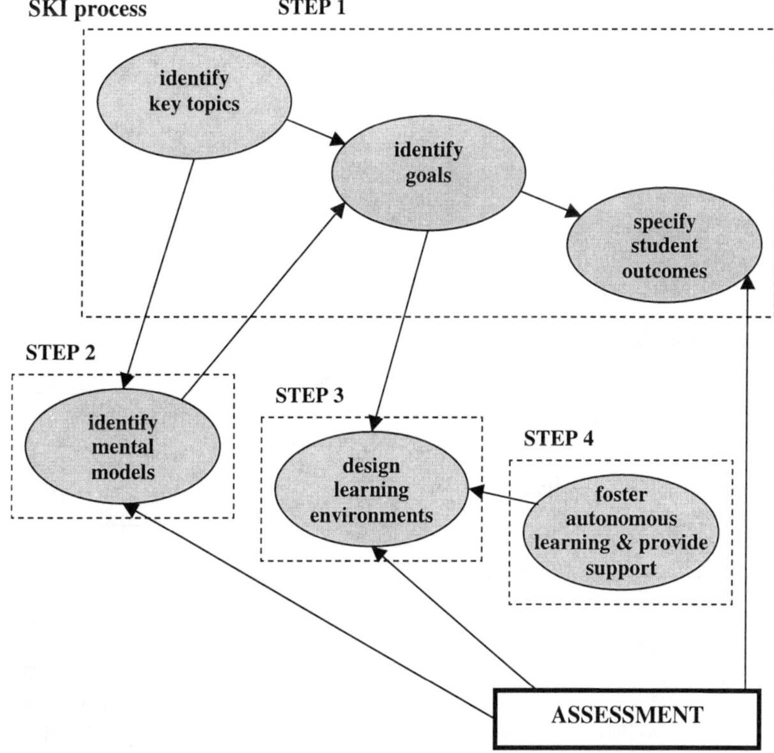

Figure 2. Schematic of Designing a Course Following the Scaffolded Knowledge Integration (SKI) Framework.

Table 1. Identifying key topics, selected instructional goals (Step 1) and corresponding models (Step 2) of students "before instruction" and "after instruction", and of experts in soil mechanics.

Step 1:	Identify key topics and goals	Step 2:	Identify	models
Key topic	*Selected instructional goals*	*Student "before"*	*Student "after"*	*Expert*
[a]Solids-pore fluid interaction	Anticipate behavior	X	X	
	Describe & explain effects		X	X
[a]Response of solids and water to loading	Anticipate trends	X		
	Use mechanical models for explanations		X	X
[a,b]Effective stress	Explain using analogies		X	X
	Define in own words		X	X
[b]Stress distribution due to loading	Develop order-of-magnitude sense	X	X	
	Predict/justify differences between 2-D and 3-D	X	X	X
[a,b] Response of soil to shear	Role of shear strength in normal stress distribution	X	X	X
[b] Link between analysis and testing	Specify tests to obtain parameters needed for analysis		X	X
[b]Limit equilibrium method	Apply method to specific problems		X	X
	Sketch application of method to different problems		X	X

[a] Qualitative description of the behavior of soils
[b] Quantitative analysis of the behavior of soils

Example key topics are listed in Table 1, together with selected instructional goals. Because there is no standard soil mechanics course, the table includes a variety of key topics to accommodate course emphasis on environmental problems, on classical mechanics, or on geotechnical engineering applications. A discussion on these key topics is included in this paper to reflect the author's belief that course design requires that instructors articulate explicitly their reasoning for topic selection, ideally in a language intelligible to colleagues outside the instructor's sub-discipline. The first key topic in Table 1, solids-pore fluid interaction, is selected because the particulate, multiphase (i.e., solids, water, air) nature of soils causes some difficulties in the conceptual, qualitative understanding of their behavior when the soil pores are filled with a fluid other than pure water. The potential for interaction between the solids and the water phase deserves emphasis in a soils class because these interactions are not typically discussed in any other undergraduate civil engineering course (with the exception of a possible elective course in geoenvironmental engineering), and are very important for subsurface contamination problems. For instructional purposes, highlighting these interactions serves as a powerful reminder of the multiphase nature of the soil. The particulate nature of soils also introduces the need to visualize the response of solids and water to loading, i.e., to appreciate the fact that both the solids and the water play a role in the load bearing capacity of the soil. These two key topics require qualitative description of the soil behavior and, as a result, the selected instructional goals are of the "know that" type, such as anticipating trends and being able to explain phenomena using analogies to equivalent mechanical models (see Table 1).

By introducing the concept of effective stress, soil mechanics is able to adapt solutions derived for continuous media. In calculating stress distribution due to loading, students need to expand their background on stress within members and structures of finite dimensions, by thinking in terms of stress distribution within a semi-infinite domain. In the same way that the students need to expand load-stress distribution concepts from previous classes, they also need to extend their ideas on the engineering behavior of materials to include failure criteria that describe the response of soil to shear.

A distinct feature of soil mechanics is the development of a number of simplified theoretical analyses for specific loading conditions that involve soil parameters determined from specialized laboratory tests designed to model similar conditions. Two unique characteristics of these analyses should be emphasized during instruction. First, the close link between analysis and testing needs to be stressed. Second, it is worth justifying the simplifying assumptions involved in these analyses by demonstrating the good agreement between the results of simplified methods and advanced numerical tools (e.g., limit equilibrium stability analysis and finite element analysis, respectively). The limit equilibrium method provides an opportunity to demonstrate this good agreement and an analytical technique suitable for a variety of geotechnical engineering problems. The instructional goals for these topics which require quantitative analyses include a blend of declarative and procedural skills, such as using analogies in explanations, developing a sense for the magnitude of the quantities involved and obtaining experience with applications. Step 1 is complete when the primary instructional goals are finalized, taking into account results from Step 2, and translated to measurable student outcomes.

Step 2 - Identify models about key topics

After the instructional goals are identified, their contents need to be specified in terms of the relevant thinking processes, or mental models (see Figure 2). These models can be identified with suitable questionnaires, annotated problem solutions and interviews from experts and students. Questionnaires and problems are being designed for the specific goals defined in Step 1; a few examples will be discussed below. The analysis of the interviews will be guided by developed protocols (McGilly 1994; Pirolli and Recker 1994). Experts typically provide a large repertoire of strategies, which may not though be the most effective for students (Linn 1995). Students' models of successful strategies may be a more appropriate goal for instruction. The outcome of Step 2 will consist of a wide set of student/expert models and explanations, which will be explicitly incorporated in instruction. Hence, students will have a variety of learning tools to choose from in order to accomplish the specified instructional outcomes. In addition, analysis of the models will clarify areas where students have difficulties and guide the design and production of instructional tools during Step 3.

A group of ten instructors of soil mechanics courses, including instructors involved in professional practice, will be the expert group for this work. The student group consists of Carnegie Mellon undergraduates enrolled in a soil mechanics course in the spring semesters of 1999 and 2000. Undergraduate students in geotechnical engineering at UC Berkeley will also be questioned during the 2000-2001 academic year to reveal any inter-institutional differences.

As discussed earlier, the key topics in soil mechanics involve models for two sets of skills: anticipating trends in soil behavior for specified conditions (i.e., qualitative skills), and selecting and carrying out correctly the appropriate analysis for these conditions (i.e., quantitative skills). The topics requiring qualitative vs. quantitative representation skills are indicated in Table 1. It is important that students' models be identified both before and after instruction. Students come to class with preconceptions (i.e., prior models) that are relevant to a varying degree to some of the key topics, indicated by "X" in the "Student before instruction" column of Table 1. With soil being such an abundant natural material, qualitative skills are very much influenced by students' models derived from intuitions and personal experiences. Quantitative models are partly based on knowledge from prerequisite courses. Exposing these models helps the student modify, refine and blend them successfully with new knowledge given during instruction. The purpose of identifying students' models following instruction is to both understand how students make sense of key topics, as well as to develop a repertoire of successful student strategies.

The models that correspond to the two sets of skills need to be accessed in different ways. To identify the models of experts regarding qualitative descriptions of soil behavior, a set of open-ended questions will be put together, such as "How do you understand the ... (e.g., loss of strength of a clay when mixed with salt water)?" or "How would you explain to a civil engineering student without a soil mechanics background the ...?" Students' ideas on qualitative behavior can be identified through "before-after-instruction" questions. The premise behind using these before-after questions is that the instructor seldom "writes on a clean slate"; the students have always some prior models, however incomplete or loosely connected, that need to be integrated with new material. The students are asked to complete the questionnaires prior to instruction, free from any expectation to "get the right answer". As the relevant material is covered, their "before-instruction" answers are being discussed in class and students are invited to modify them. The same questions can be used to obtain "after-instruction" answers as part of regular assignments, or in exams. Before-after answers provide the students with an opportunity to monitor their improvement and the instructor with an assessment tool naturally integrated with instruction.

A sequence of questions on forces exerted on a retaining wall demonstrates how students' models can first be revealed and then changed following instruction. On the first day of a soil mechanics course, 35 students were asked to consider a retaining structure embedded in solid, impermeable rock to contain (a) water and (b) dry soil, and guess in which case the wall feels a greater force. They were further instructed to provide multiple justifications for their answers by completing cues, such as "The reason I think so is..." and "I base my guess on the fact that..." (the complete assignment is included as Appendix A). The students knew that this was an exercise meant to facilitate class discussion and not to be graded. This exercise gave the following results on the models of students before any soil mechanics instruction. One student returned a blank assignment. One student answered that the answer depends on whether soil or water is heavier. Three students guessed that the wall holding back soil feels a greater force, i.e., chose case (b), and the remaining 30 chose case (a); a soil mechanics instructor would be delighted at the students' intuition. However, from the 30 students who chose case (a), only 10 students gave reasonable justifications. Eighteen of these 30 students justified their answer by writing that water is heavier than soil! (It is worth noting that only two out of the 18 questioned their belief that water is heavier.) The reasonable justifications can be summarized with sentences such as (the frequency of each answer is included in parenthesis): 1. "water can move more so it results in a higher force" (nine times), 2. "dams are larger than retaining walls" (five times), 3. "water pushes in all directions but soil mostly downwards" (once) and 4. "soil has some ability to support itself" (once). All three students who chose (b) justified their answer by saying that soil is heavier than water. This exercise motivated an all too memorable (see next paragraph!) clarification that water is not heavier than soil. It also provided opportunities for reminders that soil indeed can "hold itself up" better than water, using "what-if questions" as prompts (see Appendix A), such as "what if we remove the walls in (a) and (b)?" or "what if we open a hole in each wall?". It is important to note that these "before-instruction" questions not only help identify misconceptions early on, but also provide useful explanations in an every-day language (free of domain-specific terms) that makes sense to students. It is ironic, but perhaps inevitable, that instructors (domain experts) rarely speak this language naturally; instead, they feel more at home with the domain-specific technical terms.

After instruction on shear strength and lateral earth pressures, 28 of these same students were given in a midterm exam a problem that required of them, among other things, justified comparisons of the magnitude of the horizontal force exerted on a retaining wall for four different materials and drainage conditions. The "before question" of the first day of classes was discreetly embedded in this exam problem. Eight students answered that the soil exerts a higher force compared to water because by this time they had learned all too well that soil is heavier than water! For those students, instruction was only partially successful: it changed what students accepted as facts but not their reasoning. Twenty students answered that water exerts a higher force; of those, 17 gave the right explanation.

This sequence of the retaining wall questions was completed a few days after the midterm exam with another in-class, not-for-grade problem involving two retaining walls (a) and (b). The two walls this time held back dry soils with angles of internal friction ϕ_1 and ϕ_2, respectively, where $\phi_1 < \phi_2$. The question was again which wall feels a greater force and why. Of the 23 students who answered, 14 chose (a), nine chose (b). Reading in detail the justifications of the 14 correct answers was a delight: students gave four different arguments, using tools such as: 1. failure criteria ("the soil will fail first below the lowest failure envelope"), 2. plain English ("the stronger soil will hold itself more"), 3. visualization of the failing mass ("the lower ϕ_1 soil will have a larger failing mass pushing the wall"), and 4. sample calculations of the force for two sets on numbers. Most of the students who gave the wrong answer followed the fourth strategy: they either used the wrong formula for the active earth pressure or made a calculation error. This final question provides an example of a successful student strategy and its application: students extended further their understanding of the soil being able to hold itself up better than water (a "plain-English" strategy) to arguing that the stronger ϕ_2 soil with hold itself up more than the weaker ϕ_1 soil.

Annotated expert solutions available in the literature provide models of the thinking processes of experts (Collins et al. 1991). For the present work, annotated student solutions to selected assignments are used to obtain a more in-depth analysis of how students' models come into play during applications of basic concepts and analytical methods. (They also provide a faithful assessment tool since, as described above, students can give the right answers for the wrong reasons!) The author first introduced annotated assignments in the spring semester of 1999, primarily to test the degree to which the students could follow instructions in preparing annotated solutions. The students had to submit annotated solutions to selected problems throughout the semester, after attending in-class demonstrations of what an annotated solution may look like and receiving a set of example annotated solutions.

Several of their answers exceeded the author's expectations and augmented the example set.

One of the questions that gave insightful annotated answers was whether "we can treat a river as a strip load for the underlying soil deposit". The inspiration for this question came from a student saying "a river" when the author invited the students to give examples of vertical strip loads (the author expected "a long earth embankment" as an answer!). Instead of "correcting" the student, the author said that his answer was worth making an assignment out of it and thanked the student for the inspiration. His answer became a question in a problem set where the students also had to calculate effective stresses in a soil deposit at an offshore location. This grouping of "effective stress" problems was an indirect hint that provided scaffolding to the reasoning of the students as they moved from well-defined to open-ended questions. These annotated solutions not only gave the author a chance to "peek into the students' brains" and understand the different ways students can perceive a question and justify an answer, but also allowed the students to slow down and become aware of their own thinking and assumptions. These points are highlighted with a few student answers. One student answers yes because, according to this student, the flowing river exerts a horizontal force on the soil at the river bottom (this student obviously forgot along the way that the question dealt with a vertical strip load). One student begins by thinking that if the river is "separated" from the water table, then it could act as a load. She then goes on to say that if however she considers a picture where the surface of the river is at the same elevation as the water table in the surrounding soil, then the soil will not feel the river as a load because "the water will hold its own weight". Another student fills a page furiously computing gradients in order to calculate any change in effective stress before becoming aware that he tacitly assumes that there will be vertical flow of water in the soil.

The first pilot of annotated assignments made clear to the author that their solutions, together with clarifying insights and eye-opening misconceptions, are bound to contain "half truths". To address this, the author plans to give a smaller number of annotated assignments including some prompts and to complement them with interviews as necessary. The same assignments will be given to experts but their requested answers may be significantly shorter to increase the chances of expert participation in the project. For example, the assignments that focus on the correct application of a certain methodology may be accompanied with solutions including minimal explanations: the experts will be asked to highlight and further explain the points they think are critical for understanding. For the open-ended assignments, the experts will be asked to explain how they would approach the problem and to give only an outline of a solution. Experts will be also asked to contribute insights on how they developed their domain-specific skills (Hsi et al. 1997).

Step 3 - Making thinking visible

Step 3 consists of creating appropriate lecture, laboratory and software tools to "make visible" (Collins et al. 1991) the multiple representations of phenomena and ways to approach problems that were revealed in Step 2 and are consistent with the instructional goals of the course (see Figure 2). Providing a large repertoire of learning strategies will suit a wider range of students' abilities and learning styles (Sharp et al. 1997). It is important to emphasize that the selected content and learning goals ought to guide the choice among in-class, laboratory and computer-based learning environments. The choice may be in part guided by Laurillard's (1993) in-depth analysis of the strengths of different instruction media vis-à-vis the components of the learning process.

During instruction, alternative explanations and thought processes will be discussed explicitly. Answers similar to those the students gave when comparing forces by two different soils on a retaining wall can provide the material for structured class discussions. Such a learning environment allows the students to learn from each other and, with some prompting from the instructor, to reflect of their own learning preferences.

Processes that are difficult to communicate verbally will be illustrated through hands-on and computer-based activities. Involvement of the wider geotechnical engineering community will be necessary at this stage to incorporate past experience. The author is aware of one experienced geotechnical engineering instructor who is currently documenting the effective use of demonstration materials (Elton, 1999). Some general directions and possibilities will be sketched here.

Issues relevant to the key topics involving qualitative description of soil behavior can be highlighted with physical models. Short hands-on demonstrations appeal to students and provide an alternative to full-scale laboratory experiments (Kresta 1998). Computer-based tools will help students appreciate the factors entering in the quantitative analysis of soil behavior. The compatibility of the instructional goals of existing virtual testing software (Budhu 2000) and those of the course design will first be evaluated for this purpose. The small time investment of virtual testing will allow the instructor to combine analysis and testing in a single project and demonstrate their tight link.

New software tools will also be developed to accomplish the learning goals identified in Step 1 by allowing students to practice the thinking processes identified in Step 2 (i.e., the "contents" of the instructional goals). Students' difficulties will set the priorities for the development of these tools. For instance, students typically need a variety of learning activities to develop a sense of how small clay particles are and of how stresses created by a surface load attenuate in an elastic half space. Two visualization tools currently under evaluation will help speed up these cognitive processes. The techniques introduced in the popular physics software and book "Powers of Ten" (Morrison et al. 1994) can be adapted to contrast the relative magnitude of clay and sand particles. Another ongoing project evaluates the relative merits of presenting the spatial distributions of vertical and maximum shear stresses created by a surface load, compared to depicting how the same load changes the magnitude of the principal stresses and the orientation of the principal planes.

Future Work

Step 4 - Foster autonomous learning and provide support

Step 4 sharpens the focus of the instructional effort by implementing two kinds of practices that are central to the SKI framework: encouraging autonomous learning and providing supports. Autonomous learning is effected when the students are exposed to multiple representations and hence have the opportunity to compare and critique many approaches, including their own. Opportunities to exercise autonomy will be provided by incorporating in the design of the tools designed in Step 3 a variety of the domain-specific strategies identified in Step 2. The advantages of explicit discussion and practice of domain-general learning strategies, including self-explanation and self-regulation (Bielaczyc et al. 1995), will be evaluated as part of this work. The relative contribution of domain-specific vs. domain-general cognitive skills is a topic of debate; for the less procedurally rich domains, general cognitive skills may be more important than domain-specific strategies (Alexander 1992). It may be possible then that practice of general cognitive skills is more appropriate for instruction in engineering courses with less well-defined tasks, compared to the physics and computer science courses studied so far. Instructional supports can be provided by modeling thinking processes for students, designing instructional tools to incrementally guide students through activities, and by providing opportunities for the students to observe other learners with varying degrees of skills (Collins et al. 1991).

Assessment

In the methodology proposed herein, some assessment tools follow naturally from the preparatory work for the course design and can be integrated with instruction, as has been recommended recently (Ewell 1998). Assessment will focus on four measures. First, the students' performance as a group will be evaluated with reference to the instructional goals outlined in Table 1. Second, each individual student's progress will be monitored with the help of the "before-after instruction" questions. Thirdly, the students' satisfaction with the instructional tools developed and their awareness of learning strategies will be assessed by designing questionnaires based on published examples (Ko and Hayes 1994). Lastly, transferring of skills will be evaluated using questionnaires in courses that build on Soil Mechanics.

Concluding Remarks

The goal of the work-in-progress described herein is to redesign an undergraduate civil engineering course in soil mechanics to make learning an explicit part of the course. There are no new fundamental principles behind this approach, as teachers have been following these principles intuitively in modifying their teaching strategies. What is new, is changing instruction systematically within a cognitive science framework and using technology as necessary to enable these changes in a traditional engineering area. The rationale, methodology and some of the instructional tools developed will be readily used for the design of courses that build onto soil mechanics, such as geotechnical and geoenvironmental engineering. Perhaps what is more important, the educational philosophy underlying this work will help instructors shift the focus to how students actually think, from what the students ought to think.

Acknowledgments: The author had two great classes of juniors enrolled in Soil Mechanics (Course No: 12-335) in the spring semesters of 1999 and 2000 at the department of Civil and Environmental Engineering of Carnegie Mellon University; she is grateful for all that they taught her. The teaching assistant for the 2000 class, Ioannis Epanomeritakis, provided an intelligent refresher of the theory of Elasticity (and figure-drawing expertise). Laura Lee lent me her eyes to see the power of the "Powers of Ten". Indira Nair offered genuine encouragement and thoughtful feedback. Paul Steif shared generously his experience with instructional software design. Terre Cinkovic provided assistance in the preparation of the manuscript. This work will be completed with support by the National Science Foundation POWRE Grant CMS-9973358.

References

Alexander, P.A. (1992). Domain Knowledge: Evolving Themes and Emerging Concerns, *Educational Psychologist*, 27:1:33-51.

Bielaczyc, K., Pirolli, P.L., and Brown, A.L. (1995). Training in Self-Explanation and Self-Regulation Strategies: Investigating the Effects of Knowledge Acquisition Activities on Problem Solving, *Cognition and Instruction*, 13:2:221-252.

Budhu, M. (2000). A Virtual Triaxial Test Courseware, this volume.

Chi, M.T.H, Feltovich, P.J. and Glaser, R. (1981). Categorization and Representation of Physics Problems by Experts and Novices, *Cognitive Science*, 5:2:121-152.

Collins, A., Brown, J.S., and Holum, A. (1991). Cognitive Apprenticeship: Making Thinking Visible, *Am. Educator*, 15:3:6-11, 38-46.

Elton, D.J. (1999). E-mail communication on "Geomagic" (to appear as Geotechnical Special Publication).

Ewell, P.T. (1998). National Trends in Assessing Student Learning, *J. of Engineering Education*, 87:2:107-113.

Hsi, S., Linn, M.C., and Bell, J.E. (1997). The Role of Spatial Reasoning in Engineering and the Design of Spatial Instruction, *J. of Engineering Education*, 86:2:151-158.

Ko, E.I., and Hayes, J.R. (1994). Teaching Awareness of Problem-Solving Skills to Engineering Freshmen, *J. of Engineering Education*, 83:4:331-335.

Kresta, S.M. (1998). Hands-on Demonstrations: An Alternative to Full Scale Lab Experiments, *J. of Engineering Education*, 87:1:7-9.

Laurillard, D. (1993). *Rethinking University Teaching: a Framework for the Effective Use of Educational Technology*, Routledge, New York, NY.

Linn, M.C. (1995). Designing Computer Learning Environments for Engineering and Computer Science: The Scaffolded Knowledge Integration Framework, *J. of Science Education and Technology*, 4:2:103-126.

McGilly, K. (1994). Cognitive Science and Educational Practice: An Introduction, in *Classroom Lessons: Integrating Cognitive Theory and Classroom Practice*, K. McGilly (Ed.), MIT Press, Cambridge, MA.

Morrison, P., Morrison, P., and the office of C. and R. Eames (1994). *Powers of Ten: About the Relative Size of Things in the Universe*, Scientific American Library, New York, NY.

Pirolli, P. and Recker, M. (1994). Learning Strategies and Transfer in the Domain of Programming, *Cognition and Instruction*, 12:3:235-275.

Reif, F. (1995). Millikan Lecture 1994: Understanding and Teaching Important Scientific Thought Processes, *Am. J. of Physics*, 63:1:17-32.

Sharp, J.E., Harb, J.N., and Terry, R.E. (1997). Combining Kolb Learning Styles and Writing to Learn in Engineering Classes, *J. of Engineering Education*, 86:2:93-101.

Svinicki, M.D. (1991). Practical Implications of Cognitive Theories, in *College Teaching: From Theory to Practice*, R.J. Menges and M.D. Svinicki (Eds.), Jossey-Bass, Inc., San Francisco, CA.

APPENDIX A

"BEFORE 12-335 #1"

January 11, 1999

A retaining structure is embedded in solid, impermeable bedrock to contain (a) water and (b) dry soil, as shown in the sketch below. In which case does the retaining wall feel a greater force?

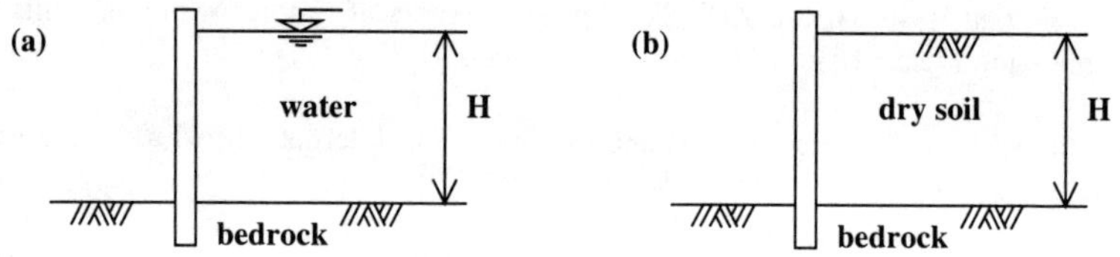

What?

A. My guess is that the wall feels a greater force in case …

Why A? - arguments

B1. The reason I think so is…

B2. I base my guess on the fact that…

B3. …

Why A? - calculations

B4. If one were to base the answer to A on some calculations, one would need information on the following characteristic(s) of water/soil/…

Why (not) A? - improve, strengthen, modify "why A" arguments

C1. My argument is not very convincing because…

C2. A better argument would be…

C3. An important point I haven't considered is…

C4. What if … (modify problem)

What? - do you want to reconsider your answer to A?

Subject Index

Page number refers to the first page of paper

Author Index

Page number refers to the first page of paper